Mathematical Research

Computer Analysis of Images and Patterns

edited by

K. Voss
D. Chetverikov
G. Sommer

Volume 55

AKADEMIE-VERLAG BERLIN

In this series original contributions of mathematical research in all fields are contained, such as

— research monographs
— collections of papers to a single topic
— reports on congresses of exceptional interest for mathematical research.

This series is aimed at promoting quick information and communication between mathematicians of the various special branches.

In diese Reihe werden Originalbeiträge zu allen Gebieten der mathematischen Forschung aufgenommen wie

— Forschungsmonographien
— Sammlungen von Arbeiten zu einem speziellen Thema
— Berichte von Tagungen, die für die mathematische Forschung besonders aktuell sind.

Die Reihe soll die schnelle Information und gute Kommunikation zwischen den Mathematikern der verschiedenen Fachgebiete fördern.

Manuscripts
in English and German comprising at least 100 pages and not more than 500 pages can be admitted to this series. With respect to a quick publication the manuscripts are reproduced photomechanically. Authors who are interested in this series please turn directly to the 'Akademie-Verlag'. Here you will get more detailed information about the form of the manuscripts and the modalities of publication.

Manuskripte
in englischer und deutscher Sprache, die mindestens 100 Seiten und nicht mehr als 500 Seiten umfassen, können in diese Reihe aufgenommen werden. Im Interesse einer schnellen Publikation werden die Manuskripte auf fotomechanischem Weg reproduziert. Autoren, die an der Veröffentlichung entsprechender Arbeiten in dieser Reihe interessiert sind, wenden sich bitte direkt an den Akademie-Verlag. Sie erhalten dort genauere Informationen über die Gestaltung der Manuskripte und die Modalitäten der Veröffentlichung.

Computer Analysis of Images and Patterns

Mathematical Research Mathematische Forschung

Wissenschaftliche Beiträge
herausgegeben von der
Akademie der Wissenschaften der DDR
Karl-Weierstraß-Institut für Mathematik

Band 55

Computer Analysis of
Images and Patterns

Computer Analysis of Images and Patterns

Proceedings of the III. International Conference CAIP '89 on Automatic Image Processing

held in Leipzig (GDR), September 8 – 10 – 1989
organized by Scientific-Technological Society for Measurement and Automatic Control (WGMA)

---

edited by

Klaus Voss
Dmitry Chetverikov
Gerald Sommer

Akademie-Verlag Berlin 1989

Herausgeber:

Prof. Dr. Klaus Voss
Friedrich-Schiller-Universität Jena
Sektion Technologie

Dr. Dmitry Chetverikov
Hungarian Academy of Sciences, Budapest
Computer and Automation Institute

Dr. Gerald Sommer
Friedrich-Schiller-Universität Jena
Sektion Technologie

Die Titel dieser Schriftenreihe werden vom Originalmanuskript der Autoren reproduziert.

ISBN 3-05-500755-7
ISSN 0138-3019

Erschienen im Akademie-Verlag Berlin, Leipziger Str. 3-4, Berlin, DDR-1086

Lizenznummer: 202·100/415/89
Printed in the German Democratic Republic
Gesamtherstellung: VEB Kongreß- und Werbedruck, Oberlungwitz, DDR-9273
Lektor: Dipl.-Phys. Gisela Lagowitz
LSV 1085
Bestellnummer: 764 071 8 (2182/55)
03500

P R E F A C E

This volume is a collection of selected papers submitted to the 3rd International Conference on Computer Analysis of Images and Patterns - C A I P ' 8 9 - held in Leipzig, GDR, September 8 - 10, 1989.

The succes of the first conference 1985 in Berlin and the second conference 1987 in Wismar has encouraged the Image Processing Group in the Scientific-Technological Society for Measurement and Automatic Control (WGMA) to organize a further conference and to edite this proceeding volume.

About one third of all submitted papers are arranged in the proceedings.The selection of papers for proceedings, lecturers or posters was done by an international scientific program committee. The final arrangement of this volume was taken by K. Voss and G. Sommer in cooperation with the Executive Board of the Image Processing Group in WGMA especially with Prof. Klette.

In preparing this book, we have asked the members of Program Committee to present their meanings and opinions, their hopes and wishes in our common field of research in short papers to the community of image processing enthusiasts. Maybe these papers can influence and stimulate the discussions about image processing and the further development of image processing.

Our thanks are due to members of Scientific Program Committee, to Professor Klette as Chairman of the conference, to Dr. Englisch for his organizational work to host CAIP ' 89 in Leipzig, to Mrs. Bahn and Mr. Müller from WGMA for their efforts to ensure a successful conference, to Mrs. Lagowitz and Dr. Höppner from Akademie-Verlag for the fruitful cowork in editing this proceeding volume.

Klaus Voss

Dimitry Chetverikov

Gerald Sommer

Table of Content

CAIP'89 – Scientific Program

Author Index

COMPUTER VISION: GOALS, MEANS, AND PROBLEMS

Dmitry Chetverikov

Computer and Automation Institute,
Hungarian Academy of Sciences
Budapest, P.O.Box 63
H-1502 Hungary

Abstract: In this short note, the author presents in an informal way his personal views on some general problems facing computer vision. The main issues discussed are the ultimate goal of computer vision, computer versus human vision, heuristic versus theory, and significance and verification of experiments. The author responds to the recent papers by R. Haralick (1) and K. Price (2) who dwell on similar problems, agrees with them on some points and argues about others.

1. Computer vision as a science.

According to Haralick (1), the theory of computer vision is expected to provide "laws and principles by which computer algorithms can be designed to solve a variety of vision tasks", including general 3D scene understanding. These laws and principles are to be mathematically derived from the initial problem statement, assumptions about the particular vision phenomenon considered, and the general laws of computer vision. The aim of the experiment is either to obtain data that would hopefully facilitate the derivation of a theory concerning the phenomenon being studied, or to test an existing theory. A report on an experimental study in computer vision is expected to contain "clear descriptions of controlled situations under which the experiments are performed, a precise statement of the algorithm being used, and a statement of the results which includes some measure of the certainty of the stated results".

To this clear statement of requirements we can only add that like any serious theory, the theory of computer vision must have a predictive power, i.e. be able to predict the behavior of a vision algorithm in given circumstances. Experimental results must be general enough, statistically significant, and reproducible.

In this paper (1), Haralick concludes that presently computer vision fails to meet these basic requirements any hard science is expected to meet. Price (2) comes to a similar conclusion. Here are some of the main problems facing computer vision.

- Computer vision lacks clear and consistent conceptual basis and terminology.
- There is a gap between the theory and the practice of computer vision; heuristic and ad hoc methods dominate.
- In most cases, no theoretical criteria exist to evaluate the performance of vision algorithms.
- Often, there is no way to automatically set the optimal values for the parameters of the algorithms.
- A lot of experimental results are statistically insignificant and irreproducible.
- It is very difficult to share computer vision software; most of us prefer implementing our own programs instead of using the existing ones.

- In general, we pay too little attention to each other's results; our efforts are separate and unrelated.

Both Haralick and Price are dissatisfied with the present situation and suggest certain remedies. Before any attempt to change the situation is made, it is necessary to agree upon the ultimate goal of computer vision, define the object of research, and clarify basic concepts and terminology. In the rest of this note, we will discuss the first two of these issues. We will also touch some of the above mentioned problems.

2. The ultimate goal of computer vision.

Computer vision aims at studying and creating artificial vision systems. "Studying" means understanding the mechanisms of vision, while "creating" means developing systems for solving particular vision tasks. These two sorts of scientific activity must cohere. Currently, they seem to be pursuing different goals. Those who "study" often do not bother with the feasibility of their methods, while those who "create" pay little attention to understanding. There are several reasons for this. One of them is the engineering versus scientific mentality dualism typical for research people in general. Another one is the "application pressure" and the extensive character of the development of this immature field of science.

Still there is a deeper reason lying in different attitudes toward the aim of computer vision. It is usually believed that the ultimate goal of computer vision is to duplicate the vision capabilities of humans. This pursuit is different from the one that aims at developing a theory to support the design of systems that solve vision tasks. The key question is whether one agrees that human vision is superior in all cases and that duplicating such sophisticated capabilities as vision, intuition etc. is possible. If not, emulating human vision is not, in the long run, a valid pursuit. We must keep in mind that any electronic device, software etc. aimed at artificial vision will differ in operation and performance from human vision.

Personally, I do not believe in the possibility of the duplication, and I do not think that the object of the computer vision research is human vision. I do not share the purely pragmatic, application-centered attitude either. Most of us want to build systems that solve real problems with real input data, but also want to understand things. I do not believe we want computer vision to be a collection of ad hoc tricks (kept in secret by companies, the military, etc.).

We study physical reality with the help of vision sensors and devices and mathematical structures capable of processing and interpreting, on a high level of abstraction, information provided by the sensors. The aim is to teach such devices to "see", i.e. solve sophisticated vision problems. Trying to reach this aim we learn from human vision. The computer vision research and the human vision research profit from each other. However, there is no guarantee that the mechanisms that are used by human vision are useful for computer vision also. For example, it has been discovered that in the spontaneous texture perception by humans, the second order image statistics play a dominant role (3). This mechanism is effectively used in computer vision also. On the other hand, the discovery of textons (4) - intensity features which are the structural elements of human texture perception - seems to have no impact on texture analysis by computer. The crucial point here is the possibility of extracting the necessary features from the image.

The ultimate goal of computer vision is <u>not</u> to compete with human vision, but to develop a coherent theory facilitating the design of hardware and software to solve vision tasks. Pursuing a wrong goal we will sooner or later find ourself in the situation familiar to the artificial intelligence community: the "theoreticians" will reject any working vision system as not being "the true artificial vision", while the "pragmatists" will do the same because the system does not solve their particular tasks.

3. On some problems.

What tasks do we mean when we speak about a theory supporting the solution of vision tasks? Haralick (1) emphasizes that more effort should be put into "the definition of canonical computer vision subproblems and ... their solution". Without the definition of canonical subproblems developing a computer vision theory is hardly possible. The set of canonical subproblems would reflect the present state-of-the-art. It could be updated when new ideas enter the field.

Some of the canonical subproblems are well-defined, some (e.g. edge detection) are not. It may be argued that any well-defined subproblem has nothing to do with image analysis unless an adequate image model is given (Which rarely happens). Or the statement of a subproblem assumes that certain information is provided (To provide it is another, perhaps even harder problem). I agree that life is not easy. But there is no other way.

Haralick (1) insists that computer vision problems must be solved under some criteria of optimality. He argues that unless a problem is stated as some kind of optimization problem one does not really know how robust a solution to this problem is. I think that in computer vision research, this aim is extremely difficult to reach since quite a few problems seem to loose their "original" meaning when expressed as optimization problem. Optimality is very much goal dependent and this may be misleading, especially when the underlying assumptions are irrealistic. What shall I do with, say, those "optimal" threshold selection techniques that fail in practice? Anyway, we should try to give clear statement of the problem, the assumptions, the proposed methods, and the results. Stating the problem in an optimization manner would help one judge upon the applicability of a solution. It could also facilitate the automatic setting of parameters of vision algorithms which I consider to be one of the most important goals.

Price (2) argues that simple problems must not be solved with too complex methods. If a practical solution exists outside the realm of computer vision, this solution must be applied. A classic example of such a situation is the so called "bin picking" task in the industrial machine vision. Why to throw the parts into a bin before picking them by a robot? Obviously, as an industrial application problem this task is nonsense since the industrial environment is highly organized (Well, not everywhere). However, it could make sense in a different environment. Designing a vision system capable of handling multiple overlapping 3D objects is a challenging scientific problem.

It is common knowledge that the generality and the certainty of experimental results in computer vision suffer very often from insufficient image data and the lack of common data sets for the experiments. Apparently, everyone is happy with his/her own algorithm as if it were sufficient in itself. If the algorithm is intended for some particular application and produces the desired results for that application, I would not object unless the algorithm is published as a scientific result. But in most cases it is. A few "typical" images of a certain sort are selected and the parameters are properly tuned for these images. Price (2) suggests that vision algorithms be tested on more (at least six) real-world images of different origin in order to make the limits of the algorithm clearer. He implicity objects using synthesized images because "natural" images are usually subject to noise, distortions etc. in a very much different and often unpredictable way. However, using synthesized images appears to be the only direct way to control the data set. Another approach is to start with simple images of a very good quality moving gradually toward more realistic input data. In any case, an explicit description of the procedure used to set the program parameters should be given.

Finally, I would like to emphasize the importance of detailed theoretical and experimental analysis and comparison of different approaches to the same computer vision problem. The theoretical analysis is at least as difficult as desirable since the computer vision algorithms are very complicated. The experimental comparison is an excellent way of sharing experience about the real strength and limits of various methods - the experience you will hardly find in books.

Conclusion.

Computer vision is a young and immature science currently being in the period of extensive development: accumulating knowledge, trying to work out its conceptual basis and terminology, designing heuristic tools, extending the application area. I hope we will put more effort to go to the intensive period, the period of the development of coherent and unifying theories, the period of understanding. Otherwise, it will become clear that as a hard science computer vision is an "ill-posed problem".

References.

(1) R.M. Haralick, "Computer Vision Theory: The Lack Thereof", Computer Vision, Graphics, and Image Processing 36, pp. 372-386, 1986.

(2) K. Price, "Anything You Can Do, I Can Do Better (No You Can't)", ibid., pp. 387-391.

(3) B. Julesz, "Experiements in the Visual Perception of Texture", Scientific American, Vol. 232, No. 4, pp. 34-43, 1975.

(4) B. Julesz, "Textons, the elements of texture perception, and their interaction", Nature, Vol. 290, pp. 91-97, March 1981.

THE DIFFERENT SOURCES AND THE INTEGRATION IN IMAGE PROCESSING

Fuchs, Siegfried[1)]

The development of the field, named today computer vision or image pattern recognition was driven in the last 30 years by engineers, mathematicians, physicians, psychologists and such people, interested in the application of image processing machines. They have been motivated by very different aims: The creation of more powerful machines, the simulation of human behaviour and human intellect, the resolution of interesting mathematical problems, the exploitation and development of computers and the accelleration and qualification of any tasks to be solved using the human eyes. As a consequence the ways and methods were remarkable different in the early years and the used terms were determined by the application field. But what is to say about an intergration of this scientific field today? - Some integration results may be stated:

- First of all the integration between different application fields (e.g. biology, medicine, geology etc.) has directed the efforts to more general and common methods and to more methodological problems.
- The integration between people, which construct the computers on the one hand and such which construct the algorithms on the other hand has led to more efficient special architectures and programs.

But some necessary integration is not yet enough developed: In my opinion it is necessary to direct more attention to common mathematicial models in image analysis and in computer graphics and also to the technical realization of biological principles to get more efficient methods for pattern recognition. I see an open field in the use of the optical processing with the outstanding parallel processing abilities in hybrid systems.

From the historical way of the development we have to draw hints for the education in the field of computer vision. The education should be really interdisciplinar. This will be more necessary in the regular college education, beginning these days in our country, then in the postgradual education in the past. But the regular education offers more possibilities for the acquiring of the basic knowledge: A solid knowledge in discrete mathematics and topology, but also in statistics and in the theory of algorithms.

On the other hand the state of art enables an early practice in the field of image processing on simple available personal computers for the students.

I think that under these conditions we can expect a new generation of young people developping our field for the futural computer and human generation.

[1)] Prof.Dr. S. Fuchs, Informatik-Zentrum TU Dresden
Mommsenstr. 13, Dresden DDR-8027

THE COMPLEXITY OF APPLICATIONS IN COMPUTER VISION

Reinhard Klette

Academy of Sciences of DDR
Central Institute of Cybernetics and Information Processes
Kurstrasse 33, Berlin, 1086, DDR

Asked by Klaus Voss to submit a personal statement about the recent state and future progress in our field I like to formulate my basic care about the **slow increase in new applications of Computer Vision**. Here, the term "Computer Vision" is used as general descriptor of our field, i.e. solving vision problems on computers, where Image Processing, Image Understanding, Image Modelling etc. are subfields. First I will propose a characterization of a "space of complexity dimensions" as introduced in [2], then I will add some thoughts about research orientations in our field, and finally I will advertise my personal conclusion drawn from the scetched situation.

Computer Vision is a sub-discipline of Computer Science, and its value to the society is mainly measured by its contribution to operational systems. Of course, independent of this economic evaluations, Computer Vision is a fascinating field in science.

The complexity of applications in Computer Vision may be evaluated by the complexities of the **class of images** relevant to the desired application, of the **interpretation task** aspired to be solved, and of the **computer system** which may be applied. In these three cases dimensions of complexity may be listed to specify the complexity of a specific application as well as the adequate level of requirements. In fig.1 examples of these dimensions are given where the left-hand notion is close to the "zero point" of the complexity space, and the right-hand notion should suggest the concrete form of increasing complexity at the given complexity axis. The term "objects" denotes the pictorial representations of the 2D or 3D real-world structures which have to be interpreted.

To my opinion, so far operational systems in Computer Vision are only available close to the zero point in the joint complexity space of

Examples of complexity axes for classes of images

isolated objects ---> overlapping objects
one moving object ---> several moving objects
shape features sufficient ---> texture features necessary
object points locally detectable ---> global models for segmentation
no noisy data ---> several types of noise
2D real-world structures ---> 3D structures
human-made real-world structures ---> natural structures in natural environments

Examples of complexity axes for interpretation tasks

static image interpretation ---> image sequences, multiple images, stereo images
off-line interpretation ---> real-time interpretation
no storing of images ---> database of images necessary
interactive image interpretation ---> automatic interpretation
interpretation at signal level ---> interpretation using knowledge of the application area

Examples of complexity axes for computer systems

on-chip ---> desk-top system ---> main-frame based system
common equipment ---> specific components for vision
common software ---> vision software system
easy-to-use systems ---> different modules with high complexity

Fig.1: Examples of complexity axes.

classes of images and of interpretation tasks, despite the fact that the fast progress in computer systems which may be applied in research, development and operational applications reduces the constraints of available hardware or software. The main application area of vision systems, now and in the near future, is in automating measurements or inspections in industrial processes. Also, interactive vision systems for laboratory work as in biomedicine or for remote sensing are increasingly in operational use. In comparison to this reached level of operational applications, even in restricted domains it is Utopian if the existence of computer vision systems is announced which may qualitatively be compared with human vision. The propagation of this illusion was used, e.g., to support military research [4] for projects as "seeing weapons" or "seeing (autonomous) vehicles".

So far it should be clear what I have addressed by "slow increase in new applications" if we remember that active work in Computer Vision is done for more than 25 years already. Similar conclusions may be drawn by analyzing the situation in technical work by asking what results may be classified as absolutely fundamental to Computer Vision. E.g., for the task of selecting a specific method for edge detection for a given class of images, no method stands out as the best in general (although the Marr-Hildreth [3] approach based on the zero crossing of the second derivative of the intensity gradient appears promising), and edge detection remains to be a research topic, e.g. at M.I.T.

The open questions related to Computer Vision which may be characterized by tuples of complexity values within the complexity space defined in fig.1 are **objective problems**. By scientific work of the international community of Computer Vision scientists these questions will be solved. But, by improving **subjective decisions** the application of results of this work may be deepend. Much of the research efforts in Computer Vision are directed to the "higher levels" of Marr's theory [3] as, e.g., "shape from ..." - approaches, reconstruction of 3D-surfaces or research in stereo image modelling. But, for the "low-level" problems of extracting or characterizing image primitives still remain tasks to be solved for designing robust and correct solutions in application areas. I assume that the preference of bottom-up research (from low-level to high-level) would ensure improved applicability of obtained results.

Just to cite two open problems at the "bottom level": The dependence of shapes with respect to different digitizations has to be studied more in detail, e.g. if objects are digitized at different positions, with different orientations, at different scale, or with different digitization schemes. Or the detection of "circular shaped objects" in a raw image: How to decide between Hough-transform, or segmentation followed by approximation, or ... as relevant approach ? By fundamental research in Computer Vision contributions to a general **theory of image structures** have to be produced. This theory starts at the basic level of dealing with topological problems related to neighborhood structures (digital topology), includes digital geometry, 3D object modelling, parts of projective geometry etc. Altogether, the very complex task of developing such a unifying theory should be a **main concern** of the international community. In this way it should be avoidable that in general application-oriented research still get stuck within "low-level" problems.

Besides this hope that a unifying scientific language will improve productivity of Computer Vision also a similar idea is directed to Computer Vision systems used for developing application systems. Recently a diversity of relatively small companies is offering vision systems, and standards are not in sight. Typically, libraries of Computer Vision procedures are delivered in conjunction with specialized hardware components, and certain interfaces (tutorial systems, interpreter language etc.) are additionally offered for simplifying the use of the vision procedures. In this situation, the design of **software translators** between the application-oriented system user and the different vision systems is directed to the creation of standardized mechanisms based on the general theory of image structures mentioned above. Besides commercial solutions for specific vision systems also a general AI approach [1] was implemented. The basic idea is top-down construction of application-oriented software by application-oriented dialogues.

References

[1] Hesse,R. and R.Klette: Knowledge-based program synthesis for computer vision. J. New Gener. Comput. Syst. 1 (1988) 1, pp. 63 - 85.

[2] Klette,R.: Einfuehrung in Maschinensehen. AUTBILD'86/1, Wiss. Beitraege der Friedrich-Schiller-Universitat Jena, Jena, 1986.

[3] Marr,D.: Vision - A Computational Investigation into the Human Representation and Processing of Visual Information. W.H.Freeman and Co., San Francisco 1982.

[4] Zamperoni,P.: Das "boese" Auge - Bildverarbeitung un Ruestung. in: Wissenschaft und Ruestung (H.Baehren, J.Tatz, Hrsg.), Steinweg-Verlag, Braunschweig 1985, pp. 200 - 222.

IMAGE PROCESSING - SCIENCE OR TECHNOLOGY?

K. Voss [1)]

Since many years, I think about this question. Assuming that image processing is a science, we have to ask to the object and to the typical method based on few theoretical notions. But if we assume that image processing is a technology, we have to investigate existing technical systems and to construct new technical systems (image processing systems in the unity of hardware and software).

I think that image processing is both - science and technology. On the one hand, a digital image is a special data structure, and image processing is therefore a branch of information processing. Now, is informatics a science? Informatics with its problems in algorithms, complexity, recursive techniques etc. is a branch of modern mathematics. Is mathematics a science?

Image processing as a mathematical science is directed to investigate "simple" data structures like digital images and to transform digital images into other data structures like lists, trees, or grammars. The theoreticians in image processing have to research into quantitative and qualitative characteristics for such transforms. They have to define abstract objects (digital images, lists, trees etc.), and thereby they can build up a new and impressive mathematical world.

I do not fear that we will fail in this point. The number theoretical investigations in signal theory, the Mathematical Morphology by Matheron and Serra, our own theory of neighborhood structures, and the fractal geometry by Mandelbrot are examples for the fruitful influence of pictorial thinking.

But image processing is also a natural science on the other hand. How works the retina of animal or human eye? How work the eyes of insects? Wath is the reason for lateral inhibition, and what is with color vision and stereoscopic vision? How can we recognize and remember complex gestalts in a fraction of second? What are the principles of brain activity in vision processes?

All these questions shall be answered by biologists, physiologists, and psychologists. To answer these questions, the natural scientists need knowledge in informatics and mathematics. They have to build up models like neuronal networks and to simulate such structures using mighty computers. And on the other hand, their results will influence image processing in sense of a feedback.

Last but not least, image processing is a technical science. Optical devices, old and new sensors, ADC's and RAM's, computers and transputers, pipelines and arrays, memories and monitors are necessary to perform real image processing operations. To control operations, we need theoretical models, mathematical methods, knowledge in informatics, and practical experiences.

All together can give a good image processing system. But there is not "the best system of all". Assuming that optimization criterias are given, and that there is an optimization method to solve each exactly described problem. These things are not sufficient to find the best system because our practical and meaningful image processing problems are not exactly described.

---

1) Friedrich Schiller University, Dept. of Technology,
Ernst-Thaelmann-Ring 32, Jena, 6900, DDR

Each theoretical model tractable by mathematical methods is obtained by neglecting an infinity of features of reality. Each of these neglected features can be meaningless in respect of one aspect and meaningful in respect of another. Only an unbounded set of real digital images gives the description of our problem to be solved.

Taking into account these facts, what is then with image processing as exact science? The state of the art in image processing is that our knowledge and our capability are poor. We do not know how we can match two human faces in a tenth of a second, and we cannot read automatically handwritten manuscripts with handwritten improvements. Everyone of us knows similar examples.

But is this meaning about the poorness of image processing right? What is with our cars, ships, bicycles, jumbo jets and other traffic machines. Can they climb a hill or a tree? Can they help us to climb a hill or a tree? Can they give us some insights how we climb a hill or a tree?

Our traffic machines are very useful for fast motion. Nothing else. They are non-omnipotent surrogates for human beings. Also our image processing machines are very useful tools to do things which are too hard for a single man. The state of the art is that many practicable solutions have been found. And for yet more problems we will find practicable solutions in future.

That is my opionion about image processing. I see this discipline as a mathematical science, a natural science, and a technical science. In this community, each year will bring new ideas, new devices, new algorithms, and new applications. But it is useless to wait for an image processing system like a human being. There are such systems already since many thousand years - the human beings themselves.

INDUSTRIAL APPLICATION OF VISUAL SENSING SYSTEMS

Detlef Schmidt 1)

'Visual-Sensing-Systems' (VSS) mean that kind of optoelectronical systems, that realize single (or several) abilities of human vision. Both, direct recognition and cognition of human vision in its very complexivity are not accessible by technical systems for the time being. But particularly such VSS will surpass biological systems:

1. VSS can realize dimension-checking tasks with high precision
2. VSS can measure the brightness absolutely and faster and in up to 500 steps of the greyscale than humans with about 20 steps
3. VSS realize a very high processing frequency up to 10^8Hz for single pixels while biological systems are working very slowly about 10^1 Hz but for all pixels in parallel
4. VSS are runnig about 24 h/d while human beings are tired within 2-4h, if inspection is strenuous.

Everywhere, in industry, in building trade or other domains of national economics men are taken up with tasks like 1.-4. It is possible today to replace their visual abilities by technical systems. Naturally, there has to be an economic benefit like a reduction in production costs or the necessity for less hazardous or strenuous work for men. Industrial application for robot vision frequently needs more complexivity than enumerated in 1.-4. That's the reason why image processing has not reached any significant numbers of realizations up to now and less than has been expected some years ago.

In contrast to this there are many more applications of image-processing-systems used for process control, inspection and quality control, for here are many partial tasks, that need such abilities, named in 1.-4. VSS for these purposes can be divided into two categories:

---

1) D. Schmidt is a member of the department of Electrical Engineering of the Wismar Technical College
Philipp-Müller-Str. , Wismar, DDR - 2400

- VSS for high precision measurements at single points or distances

- VSS that have to solve inspection tasks without high precision but integrating about a whole area of the specimen.

These latter kind of application in inspection needs algorithms so as to classify the object as an integrity. Therefore in this case the task is not to get values of measuring but to classify every specimen in one of n quality-classes, almost by using a binary algorithm.

Image processing with its mostly big amounts of information has to aid image recognition. Image recognition itself is characterized by the fact, that the infinite amount of information belonging to the object has to be compressed in the course of optical and electronical processing to 1 bit. This last bit is the answer to the question whether the specimen corresponds to the tolerance-range of the quality class N or not. This kind of inspection is named as the "go-/go no-inspection".

It is to be noted that not only the typical features of the classes have their tolerance-ranges, but there are also uncertainties in picking up and processing the specimen-dates, so that its image needs tolerances, too. Its ranges exceed with the velocity and simplicity of the VSS. The existence of these two groups of tolerances leads to the fact that the probability of a correct classification is less than 100%, for classes of industrial objects mostly touch one the next. If it is required to find a defective specimen with the probability of 100%, the VSS has to be adjusted so strong, that it will put out some objects as defective ones, that would be declared faultless by a human inspector.

Application of such inspection systems always requires the connection with a mechanical sorter. Its velocity may limitate the velocity of the recognition system. The necessary information-reduction requires:

- The visual features used for classification of the object should be chosen favourably to reach an optimal concept of the VSS. Here a complete theory is still lacking today.

- The features needed for decision at last have to be kept up during all steps of processing. Processing just has the order to lift out them of the amount of non-relevant informations.

- It is important to realize a maximum of reduction already during picking up the image optically. Lighting has to be adapted to the special problem and also methods of optical processing may be used.

- Belonging to the character and the size of the texture-elements ("texels") the size of the picture elements has to be chosen. If resolution is very high, it results also in high capacitance of storage and more time for processing.

- To need as little storage capacitance as possible, research should be done, whether line scan processing and an iterative algorithm for finding the decision of classification could be chosen. If storage-operations of large amounts of dates can be avoided, the gain of time will be important.

Increasing requirements of the tasks need more and more steps of operations. The theory of two-dimensional signals and systems can help to solve problems in image processing by using digital filters and signal-transformations. Then the capacity of usual computers is not sufficient. For that reason in future visual sensing systems will be designed as multi-processor-systems. The organisation of this system is determined by the special demands of the application-task. While image processing systems for purposes in military need up to 10^4 -10^5 processors, in industrial application VSS with up to 10^2 processors will almost do today. Further progress is dependent on available ADC-technologies for fast velocity.

During recent years the cost of microelectronic devices steadily lowered, so that industrial VSS reached the range of economic viability. On the other hand demands for the complexivility of the tasks to be solved increase more and more. Due to this fact equipment developed by scientists all over the world enable better solutions, but also need often more expenditure.

It will be the job of engineers in future as well as today to find those VSS out of hundreds, that have been described in proceedings and papers, that solve the special task as well as necessary and cause low costs.

IMPACTS OF FAST COMPUTING ON

IMAGE AND PATTERN ANALYSIS

Jack Sklansky
University of California
Irvine, CA 92717 U.S.A.

Over the past ten years, digital computers have achieved a great increase in computing power as a result of faster components, larger memories, and parallel architectures. This technological advance has encouraged the development of algorithms for image analysis and pattern analysis that would have been impractical just a few years ago. In addition, it has led to new Monte Carlo methods for designing and evaluating computer architectures for image and pattern analysis.

The recent research of the Focused Research Program on Image Engineering at the University of California, Irvine reflects this trend. This research includes: 1) analysis of multidimensional data, 2) design of automatic classifiers, 3) image modeling, 4) image matching, and 5) image-analyzing architectures. We describe each of these research activities briefly below.

1. Analysis of multidimensional data

Among the basic questions that one often asks when analyzing multidimensional data are: Q1) Are the data related to each other; if so, how are they related? Q2) Can the dimensionality of the data be reduced; if so, how can it be reduced? To help obtain responses to Q1, we have developed a powerful software tool for mapping multidimensional data onto two-dimensional graphic displays and for finding clusters in this data [1].

To help obtain responses to Q2, we view each data point as a vector in feature space, where each component of the vector is a feature or property of an observed object. We have developed two types of algorithms that select small subsets of features from initially large sets, based on the assumption that the features will be used in a piecewise linear classifier. One of these forms is an expanded branch-and-band search which tolerates some nonmonotoncity in the classification error with respect to subset inclusion [2]. The second form -- a "genetic algorithm" -- is a model of the selection process in the natural selection of chromosomes from a fixed population [3]. Both forms of algorithms search through very large spaces for global optima. Such searches are made practical by fast computers. Our expanded branch-and-bound has yielded search reduction factors of about 10^{-2} over exhaustive search when the initial numbers of features is between 10 and 20. Our preliminary tests indicate that the genetic algorithms often achieve search reduction factors of 10^{-4} or more when the number of features is between 20 and 40.

2. Design of automatic classification

Fast computers have made practical the use of advanced clustering and search techniques for the design of multiple class classifiers. The clustering techniques reduce an initially large set of training data to a much smaller set of "prototypes" that represent clusters in the data. (An example of a prototype is the barycenter of the data points in a cluster.) The search techniques search for "Tomek pairs" -- pairs of prototypes in distinct classes whose minimum circumscribing circle contains no other prototypes. These Tomek pairs determine highly efficient and robust piecewise linear classifiers [4].

When both the number of classes and the number of features are large, fast computers, in conjunction with our Tomek-pair techniques, make practical the design of classifiers in which the number of new features at each model is restricted. In this way the average number of features extracted for classifying various objects can be much less than the dimensionality of the full feature space.

3. Image modeling

The use of two-dimensional and three-dimensional models is a major new thrust for image analysis that has been made feasible by fast computers. Models not only produce improved segmentation of images, they also yield improved means of estimating geometric dimensions of objects of interest in the images. We have used this approach for the construction of three-dimensional models of coronary arteries using x-ray images taken simultaneously from two distinct viewpoints [5].

4. Image matching

Image matching -- i.e., the matching of two images of approximately the same scene obtained from somewhat different viewpoints -- occurs frequently in navigation, military reconnaissance, and medical imaging. Because image matching usually requires extensive search, the advent of fast computers is making some image matching feasible. We have developed such a technique for imaging the coronary arteries from two successive x-ray images -- one taken before injection of contrast medium into the arteries, and the other after the injection. Since the arteries are not exactly in the same 3-D position at the times of exposure of the two images, we matched the images by a nonuniform geometric correction of one image, and then subtracting the corrected image from the other image. The matching technique is based on an iterative search for local translation and rotation of a rectangular window, followed by bilinear interpolation to obtain the full geometric correction. Our preliminary results encourage us to believe that we may achieve improved measurements of coronary arterial diameters by this technique [6].

5. Image analyzing architectures

Taking advantage of the possibility of operating several processors in parallel to achieve high-speed analysis of sequences of digital images, we devised a technique for dynamic assignment of tasks to these processors. Since image analysis often consists of a sequence of tasks -- e.g., preprocessing, edge detection, segmentation, feature extraction, classification -- the tasks may be pipelined. So we devised a task assignment strategy that suppresses bottlenecks in this pipelining. Preliminary results in Monte Carlo simulations show that substantial multiplications of image-analysis speeds are possible by this technique [7].

Concluding Remarks

In our exploitations of fast computers for image analysis, we note that computing has entered a new era in which solutions to some problems are not unique, not precise, yet reliable. For example, in matching two images, the final match may vary with the initial values of certain parameters of the algorithm. The best initial values may not be known to the user. But since the quality of the final match is high for a wide range of initial values of these parameters, the solutions achieved by the algorithm are not unique, not precise, yet reliable. This quality of solutions seems to occur frequently in our use of fast, parallel algorithms. In the absence of a better term, we refer to such solutions as "smart". Smart solutions also will occur in the use of "fuzzy logic" and "neural" computing. This ability to produce smart solutions is something that humans have, too; so perhaps this new era will lead to a subsequent era in which computers and humans will understand each other more readily.

Acknowledgments

The research described here was supported by several government agencies, industrial firms, and a nonprofit private foundation, as well as the University of California. These include the U.S. Army Research Office (Contract Nos. DAAG29-84-K-0208 and DAAL 03-88-K-01117), the U.S. Department of Defense (Grant No. DAAL 03-87-G-0008), Ford Aerospace, Interstate Electronics, and the W. M. Keck Foundation.

References

1. W. Siedlecki, K. Siedlecka, J. Sklansky, "An Overview of Mapping Techniques for Exploratory Pattern Analysis", Pattern Recognition, Vol. 21, No. 5, 1988.

2. I. Foroutan, J. Sklansky, "Feature Selection for Automatic Classification of Non-Gaussian Data", IEEE Transactions on Systems, Man and Cybernetics, Vol. SMC-17, No. 2, March/April 1987, pp. 187-198.

3. W. Siedlecki, J. Sklansky, "On Automatic Feature Selection", International Journal of Pattern Recognition and Artificial Intelligence, Vol. 2, No. 2, June 1988, pp. 197-220.

4. Y. T. Park, J. Sklansky, "The Use of Tomek Links in the Design of Piecewise Linear Classifiers", Sixth Army Conference on Applied Mathematics and Computing, Boulder, Colorado, May/June 1988.

5. K. Kitamura, J. M. Tobis, J. Sklansky, "Estimating the 3-D Skeleton and Transverse Areas of Coronary Arteries from Biplane Angiograms", IEEE Transactions on Medical Imaging, September 1988, pp. 173-187.

6. L. Tran, J. Sklansky, "Flexible Mask Subtraction for Digital Angiography", Hybrid Image and Signal Processing, ed. by D. P. Casasent and A. G. Tescher, Proceedings of SPIE, Vol. 939, April 1988, pp. 203-221.

7. Y. Moon, N. Bagherzadeh, J. Sklansky, "Macropipelined Multicomputer Systems for Image Analysis", SPIE Conference on Optics, Electro-Optics, and Laser Applications in Science and Engineering, Los Angeles, California, January 1989.

INFORMATION PROCESSING FORMALISMS IN PATTERN RECOGNITION AND IMAGE ANALYSIS

Yuri I. Zhuravlev [1)]

Let us examine representation, transformation and analysis of information models which provide the solution of such information processing problems for which corresponding algorithmic procedures are unknown. The procedures should be synthesized on the basis of learning. These models serve as the foundation of modern mathematical computer science technique and their definition and application is the essence of computer science as a science. Pragmatic value of computer science basic models consists in the fact that they serve as the methodological and mathematical basis of automatization and regularization of information transformation and analysis algorithms synthesis, specifically in artificial intelligence problems and those of pattern recognition and image analysis. From the point of view of the contents computer science models are characterized by certain duality: first, they set up formalisms for information representation in algorithmical knowledge bases of computer information processing systems; second, they serve as a means of systematization of information transformation and analysis algorithms in computer information processing systems, in particular in expert systems.

For over 50 years synthesis methods exercised through information transformation and analysis algorithmic procedures learning have been, conspicuously or unconspicuously, the core of mathematical range of problems of the algorithmic theory, cybernetics and, lately, computer science. Problems requiring the use of such methods emerge in the course of computer processing and transformation of structures formed from symbols, i.e. structures representing, for example, in artificial intelligence programmes, knowledge on the subject field on the whole and knowledge on a concrete problem.

The essential task of pattern recognition is construction of effective computational means on the basis of systematic theoretical and experimental research with the aim of referring formalized situation and object descriptions to corresponding classes. Such referring (recognition, classification) is based on some aggregate estimate of the situation proceeding from its description. If correspondence between equivalence classes given on a set of solutions and a set of recognition objects (situations) is established automatization of recognition procedures is becoming an element of automatization of decision making procedures.

---

1) Computer Centre of the USSR Academy of Sciences USSR 117967 Moscow GSP-1 Vavilov str. 40

In essence, pattern recognition problems are discrete analogues of problems of a search for optimal decisions. They include a broad class of problems in which it is necessary to determine on the basis of some, usually rather heterogeneous, probably incomplete, fuzzy, distorted and indirect information given, if the situations (objects, phenomena), which are rather complex, under study possess a fixed finite set of properties allowing to refer them to a certain class - problems of recognition and classification. Or it is necessary to find out, relying on similar information on a finite set of processe of the same type, to which domain of a finite number of domains these processeswill belong after a certain period of time. Setting up and solution of the problems of such type is the base of technical (including non-destructive testing) and medical diagnostics, geological forecasting (in particular, reconstruction of geophysical fields), forecasting of the properties of chemical compounds, alloys and new materials, recognition and characterization of the properties of dynamic and static objects in a complex background environment and with active and passive noise through images received with the help of various instruments and sensors, forecasting the course of construction of large-scale objects, processing of the data of natural resources remote sensing, forest-fire locating, production processes control (forecasting of fast processes characteristics values entering in critical domains).

Practical problems to solve which it is expedient to apply recognition methods are characterized by a number of specific features:

1) These are information processing problems and their solving is implemented by means of applying a certain transformations system to the available initial data. Generally this process consists of two basic stages: a) reducing the initial data to some standard recognizable form - formalized situation (object) description synthesis on the basis of the available heterogeneous information (empiric data, measurment results, knowledge on logical aspects of the phenomena (processes), information on the structure, designated purpose and operational performance (probably, the planned performance) of the object, expert data, the available a priori semantic and syntactic information); b) recognition proper - transformation of some formalized description into a standardized result matrix corresponding the selection of some finite set (fixed) of possibilities (classifying a situation (an object) among a certain class) as the result (classified solution).

2) In these problems it is possible to introduce a concept of some similarity between objects (situations), or rather between their descriptions, to formulate a generalized concept of proximity as the basis for including situations (objects) into one and the same class or different classes.

3) In these problems it is possible to operate on a certain set of precedents - examples, the classification of which is known (as far as

the problem being solved is concerned) and which (in the form of standard formalized descriptions) can be presented to the recognition algorithm for adjusting to the problem in the process of learning.

4) It is difficult to develop formal theories for these problems and to apply classical mathematical techniques, since in the situations where they arise either of the two following cases are available: a) the level of formalization of a corresponding subject domain and/or the available information are such that they cannot serve as the base for the synthesis of a mathematical models meeting classical mathematical or mathematical/physical standards and allowing investigation through classical analytical or numerical techniques; b) in principle, a mathematical model can be constructed but its synthesis or investigation are so costly (acquisition of necessary information, computational resources, time) that the costs greatly exceed the benefits of solution sought, or go far beyond the limits of existing technical means, or make the solution of the problem simply irrational.

5) These problems, by definition, have "poor" initial information which is characteristic of a complex, in terms of semantics and structure, situation (an object in some environment) - it is limited, incomplete (with skips), heterogeneous, indirect (characteristics of manifestations of the process, which can be optional and do not necessarily belong to the principal features of the mechanism underlying this process) fuzzy, ambiguous, probabalistic. On the whole, those are problems where too little is known to make the utilization of classical methods of solving (of models) possible or expedient, but still it is enough to make the solution feasible.

The goal of an algebraic approach to pattern recognition is determination of an algorithm providing the selection of all useful information from the initial data and construction the solution corresponding, completely and precisely to the information content of the data. Such solution is characterized by a minimal (relative) computational complexity, noise in the initial data, stability and statistical reliability. In the process of solving active use is made of the precendency principle, formalization of the generalized proximity concept, automatization of the algorithm adjustment to the problem, in particular, automatization of the selection of an algorithm class which is optimal for the class of problems under consideration, as well as of the principle of correction of the final decision by means of expanding the basic set of recognition algorithm models, which were used in the course decision process.

The process of problem solving is multilevel. A heuristic algorithm model reflecting the specifity of the problem is constructed at the first stage. The next stage is characterized by the work with algorithm family models generated on the basis of the principle which is chosen in a heuristic, standard way. At this stage the optimization of recognition algorithm is implemented within the framework of individual models. At the third stage the algorithm sought is synthesized from algorithms belonging to different models.

Thus, the Algebraic Approach to information processing in pattern recognition and forecasting problems provides realization of ideology allowing to synthesize an algorithm, which solves a concrete problem precisely if certain non-rigid and checkable conditions are satisfied. It is a sort of methodology of pattern recognition algorithm synthesis automatization CAD for pattern recognition and forecasting algorithms which makes it possible to analyse the presented problem previously, to consider its specific features and after this to choose the method of solving and to suggest a corresponding algorithm on the basis of it. The principal distinction of ahe Algebraic Approach from other methods consists in the fact that stage 3 is not available in them and, consequently, it is not really possible to get a precise decision. The essence of this stage is the following: it helps to overcome difficulties emerging at stage 2 (inaccuracy of heuristic recognition algorithm models, complicacies emerging during the realization of optimization in multiparametric space) and allows to get an absolutely precise, in the above mentioned sense , solution as distinct from stage 2, where, as a rule, solutions are local-extremal.

For the work with images the so called Descriptive Approach has been formulated within framework of the Algebraic Approach to information processing in pattern recognition and forecasting problems. It envisages solving of the problems connected with getting a formal image description as a recognition object and forming and choosing recognition procedures by means of investigating the internal construction, structure and content of the image as a result of the operations with the help of which an image can be constructed from sub-images, fragments and other objects of a more primitive nature, i.e. primitives, tokens and objects extracted on an image at different stages of its processing (depending on morphological and scale level according to which an image model is formed). Inasmuch as this way of image characterization is operational the whole process of image processing and recognition, including the synthesis of formal description - an image model, is considered as the realization on an image of a certain system transformations which are determined on equivalence classes representing ensembles of allowable images. Thus, in the process of recognition a hierarchy of formal image descriptions is used, i.e. image models relating to different morphological and scale levels of representation - multilevel models allowing to choose and change the necessary degree of details in recognition object descriptions in the process of recognition.

The utilization of Algebraic Approach methodology allows to develop computer recognition systems with regard for specific features of initial data, as well as for possibilities of the available computer hardware and measurment devices or requirements to them.

FAST RUNNING ORDERING AND MAX/MIN SELECTION ALGORITHMS

I.Pitas
University of Thessaloniki
Department of Electrical Engineering
Thessaloniki 54006
GREECE

Abstract

Order statistics are used in a variety of filtering techniques (e.g. median, α-trimmed mean, nonlinear order statistics filtering, morphological filtering). Their computation is relatively fast, because it requires only comparisons. This paper presents algorithms that require a significantly smaller number of comparisons and are significantly faster than the traditional approach to the order statistics filtering. It also proposes new filter structures for order statistics filtering that are much faster than the known sorting structures.

1. FAST ALGORITHMS AND STRUCTURES FOR RUNNING MAX/MIN SELECTION

The problem of running max/min selection can be formulated as follows. Let x_i, $i \in Z$ be an one-dimensional discrete signal. The output of a max/min selection filter is a sequence y_i, $i \in Z$, satisfying the following relation:

$$y_i \triangleq T(x_i, \ldots, x_{i-n+1}) \tag{1}$$

where n is the filter length and T is either the max or the min operator. Definition (1) is slightly different than the traditional one:

$$y_i' \triangleq T(x_{i-\left\lfloor \frac{n}{2} \right\rfloor+1}, \ldots, x_{i+\left\lfloor \frac{n}{2} \right\rfloor-1}) \tag{2}$$

where $\left\lfloor \frac{n}{2} \right\rfloor$ is the first integer greater than n/2 (for n odd).
However, (1) will be used for simplicity reasons, since $y_i = y'_{i-\left\lfloor \frac{n}{2} \right\rfloor+1}$.

The problem is to construct a fast algorithm for the calculation of (1). From now on, we shall assume that $n=2^k$, unless otherwise stated. This restriction, which reminds one of the FFT algorithm, is very important for the construction of a fast algorithm, as it will be seen later on. The method which will be used for the algorithm construction is the well-known "divide-and-conquer" method [5].
It can be easily seen that:

$$y_i = T(x_i, \ldots, x_{i-n+1}) = T\left[T(x_i, \ldots, x_{i-\frac{n}{2}+1}), T(x_{i-\frac{n}{2}}, \ldots, x_{i-n+1})\right] \tag{3}$$

Thus, the computation of the max/min of a sequence of n numbers can be split into the comutation of the max/min of two subsequences of n/2 numbers each. According to (3) the output $y_{i-\frac{n}{2}}$ is given by:

$$y_{i-\frac{n}{2}} = T(x_{i-\frac{n}{2}},\ldots,x_{i-\frac{n}{2}-n+1}) = T\left[T(x_{i-\frac{n}{2}},\ldots,x_{i-n+1}),T(x_{i-n},\ldots,x_{i-\frac{3n}{2}+1})\right] \quad (4)$$

Therefore the computation of $T(x_{i-\frac{n}{2}},\ldots,x_{i-n+1})$ is common in (3), (4) and can be performed only once. The output $y_{i+\frac{n}{2}}$ is also given by:

$$y_{i+\frac{n}{2}} = T(x_{i+\frac{n}{2}},\ldots,x_{i-\frac{n}{2}+1}) = T\left[T(x_{i+\frac{n}{2}},\ldots,x_{i+1}),T(x_{i},\ldots,x_{i-\frac{n}{2}+1})\right] \quad (5)$$

The computation of $T(x_i,\ldots,x_i-\frac{n}{2}+1)$ is common in (3), (5).

Therefore it can also be performed only once. In a similar way the max/min selection $T(x_i,\ldots,x_{i-\frac{n}{2}+1})$ can be further divided in the max/min computation of two subsequences $(x_i,\ldots,x_{i-\frac{n}{4}+1}),(x_{i-\frac{n}{4}},\ldots,x_{i-\frac{n}{2}+1})$, each having $\frac{n}{4}$ elements:

$$y_i^{(1)} = T\left[T(x_i,\ldots,x_{i-\frac{n}{4}+1}),\ T(x_{i-\frac{n}{4}},\ldots,x_{i-\frac{n}{2}+1})\right] \quad (6)$$

where:

$$y_i \triangleq y_i^{(0)} \triangleq T(x_i,\ldots,x_{i-n+1}) \quad (7)$$

$$y_i^{(1)} \triangleq T(x_i,\ldots,x_{i-\frac{n}{2}+1})$$

$$y_i^{(1)} \triangleq T(x_i,\ldots,x_{i-\frac{n}{2^1}+1})$$

$$y_i^{(\log_2 n-1)} \triangleq T(x_i,x_{i-1})$$

The computation of $T(x_i,\ldots,x_{i-\frac{n}{4}+1})$ is common in (6) and in the computation of $y^{(1)}_{i+\frac{n}{4}}$, whereas $T(x_{i-\frac{n}{4}},\ldots,x_{i-\frac{n}{2}+1})$ is common in the computation of (6) and of $y^{(1)}_{i-\frac{n}{4}}$. This process can be repeated until we reach subsequences of length 2. Therefore the computation of (3) is done in $\log_2 n-1$ steps. Only one extra comparison is needed in each step, namely the comparisons $T[.,.]$ in (3) and (6). All other comparisons have already been performed for the calculation of $y_j, j<i$. Therefore, the total number of comparisons per output point is given by:

$$C(n)=\log_2 n \quad (8)$$

The conventional algorithm for the max/min selection requires n-1 comparisons. Therefore the proposed algorithm is clearly superior. A diagram illustrating the algorithm for n=8 is shown in Figure 1.

(3) can be written recursively as follows:

$$y_i \triangleq y_i^{(0)}=T\left[y_i^{(1)},y^{(1)}_{i-\frac{n}{2}}\right]=T\left[T\left[y_i^{(2)},y^{(2)}_{i-\frac{n}{4}}\right],\ y^{(1)}_{i-\frac{n}{2}}\right]=\cdots=$$

$$T\left[\cdots T\left[T\left[y_i^{(\log_2 n-1)}\ ,y_{i-2}^{(\log_2 n-1)}\right],y_{i-4}^{(\log_2 n-2)}\right],\ldots,\ y^{(2)}_{i-\frac{n}{4}}\right],y^{(1)}_{i-\frac{n}{2}}\right]=$$

$$T\left[x_i,x_{i-1},y_{i-2}^{(\log_2 n-1)},y_{i-4}^{(\log_2 n-2)},\ldots,y^{(2)}_{i-\frac{n}{4}},y^{(1)}_{i-\frac{n}{2}}\right] \tag{9}$$

Therefore only the numbers $x_{i-1},y_j^{(l)}$, $1\leq l\leq\log_2 n-1$, $i-1\leq j\leq i-\frac{n}{2^l}$ must stored for the calculation of (9). The total number of registers needed is:

$$M(n)=1+\sum_{l=1}^{\log_2 n-1}\ \sum_{j=1}^{n/2^l} 1=n-1 \tag{10}$$

For example, the storage of the following numbers

$x_{i-1},y_{i-1}^{(2)},\ y_{i-2}^{(2)},\ y_{i-1}^{(1)},y_{i-2}^{(1)},\ y_{i-3}^{(1)},\ y_{i-4}^{(1)}$ is needed for the calculation of the running max/min selection according to (9), for $n=8$.

(9) leads also to a very efficient structure for the calculation of the max/min, shown in Figure 2. This structure has a throughput delay $D(n)$:

$$D(n)=\log_2 n\cdot T_c \tag{11}$$

where T_c is the time required for a comparison. All comparisons are done sequentially, therefore only one comparator can be used in the hardware implementation. The filter structure is initialized by setting all registers equal to zero at the beginning of its operation.

Another apprach to the fast running max/min selection, for arbitrary n, is the following. If y_{i-1} is known and y_i must be computed, the window is moved only by one element (x_{i-n} is deleted and x_i is inserted in the filter window). If x_i is greater than y_{i-1}, then $y_i=x_i$, i.e. the new element is the maximum. If x_i is less or equal to y_{i-1} and x_{i-n} is less than y_{i-1}, then $y_i=y_{i-1}$, i.e. the maximum is not affected by the window movement. If x_{i-n} is equal to y_{i-1} and x_i is less than y_{i-1}, the deleted element x_{i-n} is the maximum. In this case a new max selection $y_i=T(x_i,\ldots,x_{i-n+1})$ must be performed. This running algorithm is described by the following formula:

$$y_i=\begin{cases} x_i & \text{if } x_i \geq y_{i-1} \\ y_{i-1} & \text{if } x_i < y_{i-1} \text{ and } x_{i-n} < y_{i-1} \\ T(x_i,\ldots,x_{i-n+1}) & \text{if } x_i < y_{i-1} \text{ and } x_{i-n} = y_{i-1} \end{cases} \tag{12}$$

This algorithm,called MAXLINE, requires one, two and n+1 comparisons for the three cases of (21) respectively. If the input pdf is uniform, the mean number of comparisons $\bar{c}(n)$ is given by [10] :

$$\bar{C}(n)=\frac{1}{n}+\frac{(n-1)(3n-1)}{n^2}\simeq 3 \tag{13}$$

i.e for sufficiently large window lengths n, only 3 comparisons per output point are required for the computation of the running max/min selection. It is remarkable that $\bar{C}(n)$ does not increase with n.
Both algorithms described in this section can be easily extended to the two-dimensional case [10] .

2. FAST ALGORITHMS AND STRUCTURES FOR RUNNING SORTING

The output of the running sorting of a sequence x_i is a vector sequence $\underline{y_i}$ such that:

$$\underline{y_i} = S(x_i,\ldots,x_{i-n+1})$$
$$\underline{y_i}^T = \left[y_{i(1)},\ldots,y_{i(n)}\right] \qquad y_{i(j)} \leq y_{i(j+1)} \qquad 1 \leq j \leq n-1 \tag{14}$$

where S denotes the sorting operator of n numbers. The problem is to construct a fast algorithm for running sorting. This can be accomplished by using the divide and conquer method:

$$\underline{y_i} = \underline{y_i}^{(0)} = M\left[S(x_i,\ldots,x_{i-\frac{n}{2}+1}),\ S(x_{i-\frac{n}{2}},\ldots,x_{i-n+1})\right] \tag{15}$$

where $M[.,.]$ denotes merging of two sorted subsequences of length n/2 into a new sorted sequence of length n. If we denote by $\underline{y}_i^{(l)}$ vectors of length $\frac{n}{2^l}$:

$$\underline{y_i} = \underline{y}_i^{(0)} = S(x_i,\ldots,x_{i-n+1})$$
$$\underline{y}_i^{(1)} = S(x_i,\ldots,x_{i-\frac{n}{2}+1}) \tag{16}$$
$$\vdots$$
$$\underline{y}_i^{(l)} = S(x_i,\ldots,x_{i-\frac{n}{2^l}+1})$$
$$\underline{y}_i^{(\log_2 n-1)} = S(x_i,x_{i-1})$$

(15) becomes:

$$\underline{y_i} = \underline{y}_i^{(0)} = M\left[\underline{y}_i^{(1)},\ \underline{y}_{i-\frac{n}{2}}^{(1)}\right] = M\left[M\left[\ldots M\left[\underline{y}_i^{(\log_2 n-1)},\ \underline{y}_{i-2}^{(\log_2 n-1)}\right],\ \underline{y}_{i-4}^{(\log_2 n-2)}\right],\ldots,\ \underline{y}_{i-\frac{n}{4}}^{(2)}\right],\ \underline{y}_{i-\frac{n}{2}}^{(1)}\right] \tag{17}$$

(17) describes a fast algorithm for the computation of the running sorting. Since $\underline{y}_i^{(1)}$ is common in $\underline{y_i}$ and $\underline{y}_{i+\frac{n}{2}}$, and $\underline{y}_{i-\frac{n}{2}}^{(1)}$ is common in $\underline{y_i}$ and $\underline{y_i}$ and $\underline{y}_{i-\frac{n}{2}}$, each of them must be calculated only once. The only extra operations needed for the calculation of (15) are the comparisons needed for $M[.,.]$. It is known [5] that the minimal of comaprisons $C^{(0)}(\frac{n}{2})$ needed for the merging of two ordered sequences of length n/2 is:

$$C^{(0)}(\tfrac{n}{2}) = n-1 \tag{18}$$

The overall running sorting algorithm is completed in $\log_2 n$ steps. The only extra operations needed at step j, $0 \leq j \leq \log_2 n-1$ are the comparisons required in the merging of two sequence having $\frac{n}{2^{j+1}}$ elements each. Their number is given by:

$$C^{(j)}\left(\frac{n}{2^{j+1}}\right)=\frac{n}{2^j}-1 \qquad (19)$$

The total number of comparisons needed by the algorithm is given by:

$$C(n)=\sum_{j=0}^{\log_2 n-1}\left(\frac{n}{2^j}-1\right)=2(n-1)-\log_2 n \qquad (20)$$

which is of the order of magnitude of O(n). The flow diagram of the running sorting algorithm for n=8 is very similar to that of Figure 1. The only difference is that dots denote merging operations done serially, instead of comparisons. (17) can also be used to construct a computer structure which is suitable for running sorting. This structure is shown in Figure 3, for n=8. The symbol M denotes a merging circuit.

Another new running sorting algorithm, for arbitrary n, is based on the following fact: if $\underline{y}_{i-1}$ is known, y_i can be calculated as follows:

$$\underline{y}_i = (\underline{y}_{i-1} \text{ b\_del } x_{i-n}) \text{ b\_ins } x_i \qquad (21)$$

where b_del, b_ins denote deletion and insertion in a sorted list. This fact leads to a different running sorting algorithm, called SORTLINE. This algorithm uses the BINARY-DELETE and the BINARY-INSERT routines which are essentially binary search routines [5].
It is known that the binary search in a list of n numbers requires, at most, $\log_2 n$ comparisons. Therefore, the number of comparisons per output point needed by the SORTLINE algorithm in the worst case is:

$$C(n)=\lfloor \log_2 n \rfloor + \lfloor \log_2 (n-1) \rfloor +2 \simeq 2\lfloor \log_2 n \rfloor + 2 \qquad (22)$$

The two extra comparisons which appear in (22), come from the BINARY-INSERT routine, because the number to be inserted in a sorted list must be compared to the minimal and the maximal number of the sorted list, before the binary insertion routine is activated. (22) shows that the SORTLINE algorithm is by far superior than QUICK-SORT or BUBBLESORT. Both running sorting algorithms described in this section can also easily be extended to the two-dimensional case.

REFERENCES

[1] T.S.Huang (editor) "Two-dimensional digital signal processing", Springer Verlag, 1981.

[2] H.A.David "Order statistics", Wiley, 1980.

[3] J.Serra "Image analysis and mathematical morphology", Academic Press, 1982.

[4] I.Pitas, A.N.Venetsanopoulos "Nonlinear order statistic filters for image filtering and edge detection",Signal Processing, vol-10,pp.395-413, 1986.

[5] E.Horowitz, S.Sahni "Fundamentals of computer algorithms", Computer Science Press, 1978.

[6] K.Oflazer "Design and implementation of a single chip 1-d median filter",IEEE Transactions on Acoustics, Speech and Signal Processing, vol.ASSP-31,No.5,pp.1164-1168, Oct.1983.

[7] T.S.Huang,G.J.Yang,G.Y.Tang "A fast two-dimensional median filtering aglgorithm", IEEE Transactions on Acoustics, Speech and Signal Processing,vol.ASSP-27,No.1,pp. 13-18, Feb.1979.

[8] E.Ataman, V.K.Aatre,K.M.Wong "A fast method for real-time median filtering",IEEE Transactions on Acoustics,Speech and Signal Processing,vol.ASSP-28,No.4,pp.415-421, Aug.1980.

[9] J.P.Fitch,E.J.Coyle,N.G.Gallagher "Median filtering by threshold decomposition" IEEE Transactions on Acoustics,Speech and Signal Processing, vol.ASSP-32,No.6,pp. 1183-1189, Dec.1984.

[10] I.Pitas "Fast algorithms for running ordering and max/min selection algorithms", IEEE Transactions on Circuits and Systems,vol.CAS-36,No.6,June 1989.

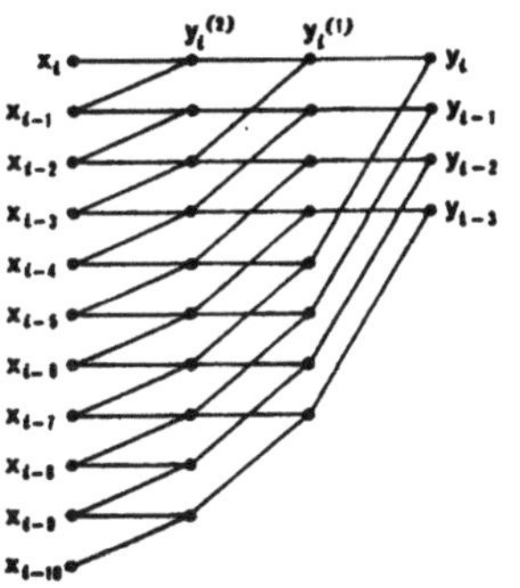

Figure 1: Flow diagram of the fast running max/min selection for n=8. Dots denote comparisons.

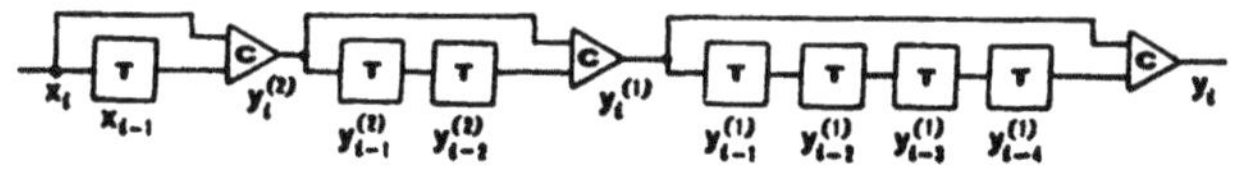

Figure 2: Stucture of the running max/min selector. C denotes comparator.

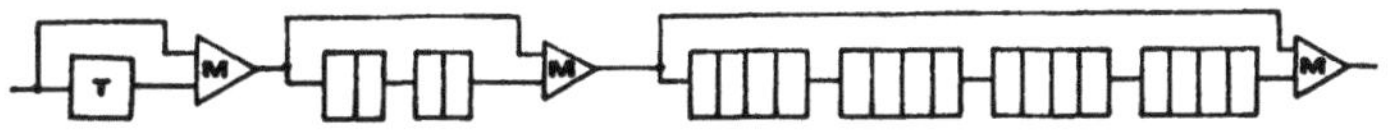

Figure 3: Structure of a running sorter of n=8 elements. M denotes a merging circuit.

IMAGE RESTORATION - AN UNIFYING APPROACH

Alexander M. Taratorin[1)]

A problem of image restoration is one of the traditional occupancies in image processing. It has originated in early sixties and is still important in many research areas (astronomy, tomography, non destructive testing and others). A lot of different algorithms for image restoration have been proposed by different authors, several monographs have been published ([1]-[3]).

In our research ([2]) we have made an attempt to look at this problem from general point of view. First of all, recorded images are always degraded by noise. The problem of restoration as an inverse problem is ill-posed, so we must use theory of regularization. At the same time, several approaches can be used to solve inverse problem: linear filtering, non-linear optimization algorithms and iterative reconstruction algorithms. We have tried to analyse common features of this approaches and their ultimate possibilities using regularization theory.

1. Linear image restoration. Standard approach to linear image restoration is to solve the following equation:

$$g = Af + n$$

where f is original image, A - distortion operator, n - noise, g is distorted image. Starting from Philips, Bacus-Gilbert, Wiener and Tikhonov regularization, many linear reconstruction approaches have been used [1,2]. They all can be explained within the framework of regularization theory as an application of linear "regularizing" operator R_α to distorted image:

$$f_\alpha = R_\alpha g$$

The typical form of R_α is as follows:

$$R_\alpha = (A^*A + \alpha Q)^{-1} A^*$$

Where α is regularization parameter, Q is some specific function. Variations of α and Q can provide some freedom in quality of restored

---

1) Institute of Radioengineering and Electronics USSR Academy of Science
103907 Marx Avenue 18 Moscow USSR

image. In general case, choice of Q provides "smoothness" of restored image and it can vary for different classes of images. Changes of α can provide a compromise between quality of restoration and level of noise in restored image (as a becomes small, noise increases). Additional degrees of freedom in controlling restoration process can be necessary to match image quality to visual perception. For example a noisy image can be preferable to noise-free but smoother image. To realize this approach additional control parameters can be added to restoration. We have proposed several controlled restoration filters, based upon degrees of inverse and smoothing frequency responses. This model can be interpreted as image restoration in distributed media.

All linear algorithms are however limited in their frequency response: they can not restore parts of the spectrum dumped to the noise level or cut off by the formation system. Because of this reason linearly restored images are always limited in resolution.

2. A class of non-linear restoration algorithms was proposed in 70-s. They were called "maximum entropy" algorithms since image restoration was considered as maximization of image "entropy" in the form of following functional:

$$\sum_n f_n \log f_n$$

Where f_n are image samples. Many variants of this methods were proposed with different functionals and all they worked with success. Enchanced restoration quality comparing with linear algorithms led to some statistical explanations of image entropy and information content. However from the numerical point of view entropy has very little relation to inverse problem. The reason of good algorithm performance is in non-negative character of solutions provided by logarithms in maximized functional. It can be shown that due to this non-linearity it is possible to expand frequency range of restored image. The main question here is how to form non-linear optimization problems: restore an image taking into consideration equations of image formation and non-negativity. We have proposed the following optimization scheme:

$$\min \{ \Phi[f] + C[f] + D[f] + \alpha \Omega[f] \}$$

where Φ is "criterium" functional (e.g. entropy), C and D are constraint functionals for equality and non-equality constraints and Ω is regularization functional providing stability of optimization to noise. Linear restoration algorithms are automatically included in this approach by using quadratic functional for Φ. Maximum entropy type algorithms are also included in this optimization scheme. At the same time it is possible to devise a new class of restoration algorithms by adding constraints to optimization problem. Numerical experiments provided in [2] proved the efficiency of this scheme for image restoration: algorithm is more stable to noise comparing with ME type algorithms and it shows high quality of restoration when additional constraints are incorporated (non-negativity, finite support, upper and lower thresholds, spectral estimates).

The main problem of image reconstruction is therefore the choice of appropriate a priori information about restored images. The only way to incorporate constraints is to use non-linear restoration algorithms.

3. Iterative reconstruction algorithms. A different approach to constraining image restoration is to use iterative algorihms. Some of them are well-known (Van Cittert, GershbergPapoulis, Fienup algorithms). A powerful theory of projections onto convex sets was developed during last decades [3]. Application of this theory to image restoration have been very successefull: many constraints can be interpreted as projections onto convex sets. In this interpretation image reconstruction can be imagined as finding and intersections of convex sets. additional constraints are making solution area smaller and better solution can be found.

However ill-posedness of solutions was not considered seriously. Noise makes the process unstable and it is necessary to develop regularization of iterative process. We have proposed several methods for introduction of regularization into iterative algorithms [2]. For example it can be shown that the set of regularized solution is convex and projection on this set can be found. In this way stability of reconstruction process can be provided easily.

The most serious problem for both iterative and non-linear reconstruction algorithms is introduction of additional constraints. For example, non-negativity constraint works well for impulse type objects on the zero background but it has little influence for realistic images. For this type of images it is necessary to use some kind of upper and lower intensity thresholds which are different for different regions of space. Such thresholds can be found from blurred image itself. Another important information is finite support constraint. A class of pseudidifferential constraint operators can be used for phase retrieval problem solution.

4. Superresolution and phase retrieval problem. After development of first non-linear algorithms it was discovered that sometimes frequency content of restored images are wider that of original image and it is possible to restore images beyond the diffraction limit. This effect was called superresolution. Theoretical studies of spectral extrapolation for analytic functions were also predicting theoretical

possibility to restore signals beyond the diffraction limit. However all theoretical predictions were greatly unstable due to noise in source data. A more practical approach to superresolution problem is based upon constrained image restoration. Non-linear constraints imposed on restored signals act as non-linear signal transformations and thus creates new frequencies. This spectral expansion is controlled by the amount of our knowledge about images (thresholds, exact knowledge of image support). It can be shown that all possible non-linear algorithms with indentical constraints are leading to the same solutions and the same type of spectral extrapolation. Practical experience shows that in case of several percent noise in input data it is possible to reach 2.5 - 3 increase of resolution if non-negativity and finite support constraints are used.

The same arguments applies to the problem of phase retrieval and image reconstruction in astronomy. Theoretical considerations about phase retrival from multidimensional spectral magnitudes are very noise sensitive. If we however include this information into general framework of image restoration by incorporating additional constraints and using the low-frequency phase information which is usually accessible in astronomical observations, problem can be solved with reasonable speed and accuracy.

It seems to us that image restoration problem can be imagined as struggle with noise and ill-posedness by using additional information abour our solution. A large work is still to be made in finding appropriate constraints and numerical algorithms to introduce them in different noise environment.

References:

1. H.C.Andrews, B.R.Hunt- Digital Image Restoration - Prentice Hall, Englewood Cliffs, NJ, 1977.
2. G.I.Vasilenko, A.M.Taratorin- Image Restoration (in Russian) - Moscow, Radio i swyaz, 1986
3. Image recovery - theory and applications /ed. by H.Stark - Academic Press, 1987.

COLOUR IMAGE SEGMENTATION BY A FUZZY DATA-FUSION APPROACH

Daniele D. Giusto, Carlo S. Regazzoni, and Gianni Vernazza

Dept. of Biophysical and Electronic Engineering
University of Genoa
Via all'Opera Pia 11A, I-16145 Genoa, Italy

ABSTRACT

This paper deals with an analysis of colour images representing outdoor natural scenes.
Some low-level processing procedures are applied, in particular, filtering and segmentation. In the filtering step, we employ a widely used edge-preserving smoothing operator, suitably modified for extension to multichromatic images.
The segmentation process is performed by a region-growing multistep operator, using both fuzzy concepts and backtracking procedures, to preserve lifelike scene descriptions.
Results concerning countryside images are reported and discussed.

INTRODUCTION

In the field of computer vision, the use of colours allows a better interaction between the human user and the machine [1], since, the human eye can distinguish thousands of colours. In the past, the possibility of utilizing colours was generally restricted to the display phase, due to the low availability or accessibility of specific hardware for the acquisition and processing of true colour images (three spectral bands: Red, Green, and Blue), and, of course, to the notable increase in complexity and computation time.
In recent years, thanks to the considerable diffusion of specific hardware at lower cost, many researchers have been able to work extensively in this area, and their findings have been reported in the literature [2, 3, 4].
New instruments have made it possible to use a larger amount of data on a given scene, thus enhancing the recognition reliability. However, some problems may arise, from the data redundancy in the three spectral bands. Many works [5, 6, 7] deal with compression of information in colour images; to attain this goal, one can use, for example, the Karhunen-Loeve transform, or various linear and non-linear techniques for combining the fundamental bands.
In the present paper, we do not consider such aspects of colour-image analysis, leaving unaltered all original information. At present, however, we are addressing also the redundancy problem, in particular, for colour-image coding.
In the following, some low-level procedures aimed at colour-image segmentation are presented; more precisely, after a section devoted to an edge-preserving smoothing operator, attention is focused on extraction of uniform regions from outdoor natural scenes, by performing appropriate steps to retain the details of their descriptions.

THE FILTERING OPERATOR

The first module of the processing system is aimed at reducing the high-frequency noise that usually affects original images, and that is generated by the acquisition process.
The traditional, basic method for reducing such noise consists in applying an averaging operator. In order to preserve edges, which may be blurred by this process, the averaging step must be performed only in uniform areas, by means of an edge-preserving smoothing operator.
Utilizing a filter widely employed for monochromatic images [8], we have developed a procedure that can be applied simultaneously to the three bands, thus carrying out a global operation.

This operator is based on the concept that, to reduce high-frequency noise, one has to calculate average values only for those regions with no contours.
Around each pixel under examination, a most homogeneous area must be found, for which to calculate average values, without degrading its contour.
For each pixel, the realized operator examines a neighbouring area, in the three bands, and averages over the most uniform (minimum variance) neighbouring sub-area, considering all three components.
The filter output is an image affected by a very low degree of high-frequency noise and characterized by a notable resolution of the region edges; this image exhibits, however, an unnatural tessellation effect, which is to be ascribed to the operator's definition.
A second kind of defect attributable to the operator lies in a considerable loss (or marked degradation) of details and filiform objects. In other words, though sharp contours strongly contribute to an accurate segmentation step, the process to obtain them may give rise to notable drawbacks, such as loss or degradation of original information.
In addition to such drawbacks, one should also consider various segmentation errors; as a consequence, a more specific processing is required in order to obtain, as a final result, an almost unaltered scene description.

THE SEGMENTATION MODULE

As a first approach to segmentation process, we have employed a conventional operator of the region-growing type [8, 9].
This operator can detect uniform regions bounded by contours that are above a (local) threshold, determined automatically by analyzing the local histogram of the differential image.
A local threshold has been computed to reduce the number of regions extracted: in fact, due to the varying image contrast, a global threshold might lead to the detection of many insignificant regions with minimum areas. By using this approach, some errors may be incurred, which will be rectified by the subsequent backtracking step. In the case of multichromatic images, more thresholds should be computed, one for each band (θr, θg, and θb).
The assignment test is carried out by considering the differences between a pixel and its 8-connected ones; possible errors caused by the presence of slowly changing areas (not detected at this stage) will be rectified by the backtracking step.
The most important modification to the segmentation process lies in the use of a membership function associated with each threshold, i.e., $Mr(\theta r)$, $Mg(\theta g)$, and $Mb(\theta b)$; these functions allow the control of "limit" situations.
Segmentation thresholds are managed according to a fuzzy criterion, and a pixel can be assigned to a region after analyzing the sum of the three membership values ($Mt=Mr+Mg+Mb$); the shape of the fuzzy membership function considered is very simple, as shown in Fig. 1.

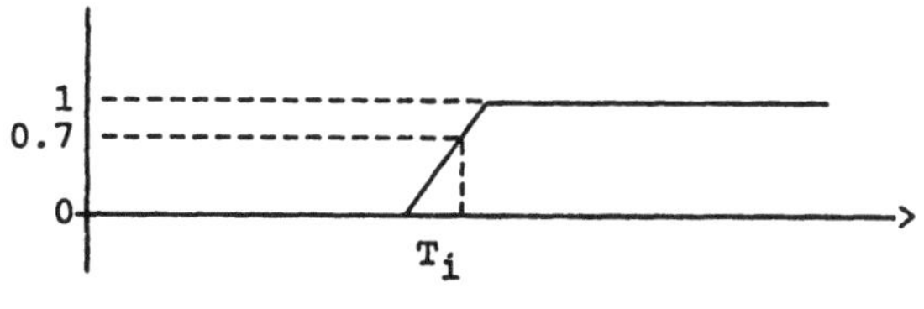

Fig. 1. The shape of the fuzzy function used for threshold computation.

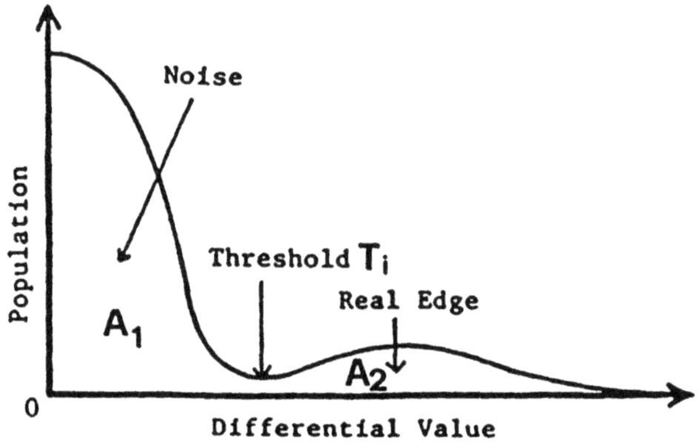

Fig. 2. The ideal case of the differential values histogram; it shows how we compute both the threshold value and the angular coefficient of the fuzzy function shown in Fig. 1.

The angular coefficient of the connecting straight line can be derived from the ratio between the areas determined by finding the threshold value on the local histogram of the differential values; for instance, if the threshold T divides the histogram into the areas A1 and A2, as can be seen in Fig. 2, the angular coefficient will be given by A1/A2.

If an analyzed area is widely uniform, the histogram of the differential values will be characterized by A1 much larger than A2, and the connecting straight line will be almost vertical. In this case, a classification can be performed without uncertainties. Instead, if an area includes many contours of various resolution values, the straight line will tend to be more horizontal, with a higher degree of uncertainty.

In conclusion, a pixel is assigned to an existing region if Mt is less than 1, and to a new region if Mt is greater than 2; otherwise, it is considered uncertain. In this way, the resulting image is divided into regions with large contours made up of non-assigned pixels.

At this point, a re-assignment step is required. The image is scanned again, and the uncertain pixels are assigned to the most similar adjacent region, or they generate new regions. This technique makes it possible to avoid assignment errors due to a hasty decision; unfortunately, filtering errors cannot be recovered, and segmentation errors inherent in the algorithm may still be incurred. In fact, regions with vertical rectilinear contours can often be observed, which are not present in any real scene, and which can be ascribed to a propagation of segmentation errors in slowly varying areas. To overcome these drawbacks, a third processing step can be performed, i.e., a backtracking process that utilizes both the label picture generated by the fuzzy segmentation process and the information about the original image (not degraded).

THE BACKTRACKING STEP

This step consists in a third image scanning, which allows the colour compactness of each region to be evaluated; new regions may be detected where a considerable scattering of colour levels occurs.

In other words, for each region and for each band, we analyze the shape of the related histogram; if more peaks are present, we search for valleys between them, and use the related grey values as segmentation thresholds.

On the contrary, if only one peak is present, we analyze its dispersion (σ), and re-assign those pixels which belong to the tails. Mixed situations are also considered, to manage the first (or the last) peak of the histogram (if more peaks are present).

In this way, pixels are re-examined and re-assigned to their "natural"

regions.
At the end of this procedure, all regions with small areas are eliminated (merging phase) and assigned to the adjacent regions with the most similar colour levels and with the longest contours [9], thus obtaining the final lifelike description of the image.

Fig. 3. The original image of a country scene including a paved road.

Fig. 4. Filtering result: the tessellation effect is clearly visible.

Fig. 5. Result of the first segmentation step: many detail appear degraded.

Fig. 6. Final segmentation result, after the backtracking step: the natural structure of each region has been recovered.

RESULTS

The adaptive segmentation process described in the paper has been tested on images representing countryside scenes characterized by the presence of different types of road (gravel roads, asphalt roads, etc.). In fact, we have developed the present processing modules for road detection, and they can be integrated into a more general image-understanding architecture to be used as a vision system for an autonomous land vehicle [10].
In the following, we report some results concerning the case of a paved road; Fig. 3 shows the original image utilized (512*512*24, displayed on an 8-bit frame buffer by means of a suitable look-up table). Fig. 4 gives the filtering result: one can easily observe the marks indicating noise reduction, contour enhancement, and detail degradation.
Those images have been utilized to perform the first segmentation step, using the fuzzy approach. The results obtained after re-assigning the uncertain pixels are illustrated in Fig. 5, where one can easily notice the filtering effects (the region contours appear too square).
These drawbacks have been overcome by the backtracking step (the last one) which has yielded a set of regions quite rich in natural details (Fig. 6). In fact, it is possible to notice the paved area (i.e., the road), the grassy areas on both sides of the road, and also many details of the nearby houses. It is worth noting that a pole is visible again on the left side of the scene, while it appears very degraded in the previous image.
The various processing steps have been performed on a Hewlett Packard 835TurboSRX computer (RISC architecture): filtering has required 5 seconds of CPU time, while segmentation has required about 1 minute.

CONCLUSIONS

An image-segmentation module has been described, that can be integrated into a knowledge-based scene-understanding architecture (under development in our laboratories) to be used as a vision system for an autonomous land vehicle [10].
Moreover, an adaptive multistep approach to segmentation of natural colour images has been proposed.
The images have first been treated by an edge-preserving smoothing operator to reduce the high spatial frequency noise caused by the acquisition step. This operator replaces the colour level of a pixel with the mean colour level of the most homogeneous region surrounding the pixel.
Subsequently, the images have been processed by using a segmentation algorithm based mainly on a classical region-growing technique, suitably modified to fully exploit chromatic information.
Various segmentation thresholds have then been utilized for the three colour bands, and a fuzzy membership function has been associated with each threshold in order to solve uncertain situations.
At the end of the segmentation process, a further scanning step has been performed to analyze the scattering of the colour levels in all regions and to re-assign uncertain pixels, thus obtaining "good" (natural) regions.
The resulting images evidence a correct extraction of the homogeneous regions present in the original images; such regions appear very natural, since all errors introduced by the filtering operator and by the first segmentation step have been rectified by the backtracking module.
In conclusion, the proposed approach, which involves some successive backtracking steps, has yielded quite satisfactory results, considering the particularly complex images analyzed. In fact, such images represent countryside scenes characterized by a large number of details and made up of regions that are not always separate enough from one another, and that exhibit different textures.

REFERENCES

[1] J.D.Foley and A.VanDam - Fundamentals of Interactive Computer Graphics - Addison Wesley, Reading (MA), 1984.

[2] R.Gershon - Aspects of perception and computation in color vision - CVGIP 32, pp. 244-277, 1985.

[3] F.Bumbaca and K.C.Smith - Design and implementation of a colour vision model for computer vision applications - CVGIP 39, pp. 226-245, 1987.

[4] B.A.Wandell - The synthesis and analysis of color images - IEEE Trans. on Pattern Analysis and Machine Intelligence, Vol. PAMI-9, No. 1, 1987.

[5] J.R.Kender - Saturation, hue, and normalization color: calculation, digitization effects, and use - Technical Report, Dept. of Computer Science, Carnegie Mellon University, 1976.

[6] Y.Ohta, T.Kanade, and M.Sakai - Color information for region segmentation - CVGIP 13, pp.222-241, 1980.

[7] Y.Ohta - Knowledge-Based Interpretation of Outdoor Natural Color Scenes - Pitman, Marshfield (MA), 1985.

[8] M.Nagao and T.Matsuyama - A Structural Analysis of Complex Aerial Photographs - Plenum Press, New York, 1980.

[9] D.D.Giusto and G.Vernazza - Optical flow calculation from feature point analysis through an automatic segmentation process - Signal Processing, Vol. 16, No. 1, pp. 41-51, January 1989.

[10] D.D.Giusto, C.S.Regazzoni, and G.Vernazza - "Multisensorial vision for autonomous vehicle driving" - Applications of Artificial Intelligence VII, M.M.Trivedi, Ed., Proc. SPIE, Vol. 1095, Orlando, Florida, March 27-31, 1989 (in press).

DIGITAL PROCESSING OF DYNAMIC IMAGES

Alexander M. Taratorin[1]

Image sequence analysis is becoming an attractive field of image processing due to new possibilities opened by the development of computer hardware and new research problems in science and engineering (biomedical images, remote sensing, nmon destructive testing, physics and chemistry). There are two large groups of problems associated with image sequences. One is the analysis of dynamic scenes (motion analysis, coding, recognition of moving objects etc). Another group of problems is very different and is concerned with the analysis of dynamic fields - continous spatio-temporal distributions of scalar or vector variables (temperature, electric potential, brightness etc).

This work is the review of author experience in development of numerical algorithms for dynamic image processing.

Dynamic image is a spatio-temporal signal with correlated values. In some way dynamic image is different from a single image as much as one-dimensional signal is different from an image. Spatio-temporal nature of dynamic image brings new techniques of image analysis.

1. Visualization. Typical visualization for image sequence is an animation of frames forward or backwards in time. However the visual analysis for smooth distributions is rather unusual for human perception since human vision is most adapted for selection of moving objects in the background. Alternative way of visualization is the analysis of spatio-temporal slices. Let the parametric curve $z(p)=\{x(p),y(p)\}$ is determined in spatial region. By looking at the dynamics of $z(p)$ in time we can obtain an image $b(p,t)$ which carry information about contour dynamics. This simple method is very useful for the analysis of complex spatio-temporal data: studies of wave processes, homogenuous changes of brightness along specific contours. For example, all moving structures in the image are appearing as specific patterns in appropriate spatio-temporal slices. Edges of this structures can be used for speed measurements, statistical processing of this edges curvature gives estimate of speed distribution.

Dynamic image can be also imagined as spatial field of temporal curves. Analysis and processing of this curves is a useful way of image processing. One of the most important applications of such an analysis is the determination of spatial region in which temporal behaviour is strongly correlated. This can be made by calculating the correlation

1) Institute of Radioengineering and Electronics USSR Academy of Science 103907 Marx Avenue 18 Moscow USSR

coefficient, claster analysis and dynamic segmentation (see later).

2. Preliminary processing is a transformation of the source frame sequence after which it becomes more useful for visual analysis and investigation. Informative changes of brightness in image sequences are often distorted by inhomogenous brightness distribution or temporal instability of sensor. A general form of this problem is the separation of image sequence into two signals: $f(x,y,t)=b(x,y,t)+n(x,y,t)$. For example, In many cases the registered distribution of parameter (eg. the temperature of brain or body) can be imagined as two distributions: one is stable and usially high amplitude and the second is weak and dynamical. To select this second distribution it is necessary to estimate the first one and then to subtract it from the image sequence. A very close method is the temporal derivation of image sequence. After this operation the stationary image regions are supressed and changing regions are enhanced.

3. Statistical portraits of dynamic images. Dynamic image is non-stacionary and non-homogenuous random field. To develop effective algorithms of image processing it is necessary to find adequate techniques for extraction of statistical information from image sequences. We have used several approaches for statistical description of dynamic images. One of the approaches is the study of dynamic gistograms. For example the simplest brightness gistogram of each frame in image sequence forms two-dimensional function H(J,T) where J is the brightness level, t - time instant, H is number of J level in the image. This function describes redistribution of probability density function of the image in time and it forms "statistical portrait" of specific dynamic process. Parameters of this function (average, dispersion, number of modes) and their redistribution in time characterises the dynamic process. Temporal derivative of H(J,T) can be used to select stacionary and non-stacionary behaviour of dynamic process. Even this simplest function characterizes specific properties of dynamic process. For example, if there are several homogenuous brightness regions in the image then H(J,t) is the sum of several modes. Evolution of this modes naturally describes homogenous spatio-temporal dynamics of image regions. This information can be used for dynamic image recognition, segmentation and functional mapping. Another useful statistical approaches are marking of stacionary regions, selection of regions with high spatio-temporal correlation and others.

4. Filtration of dynamic images. There are several important problems of dynamic image filtration: spatio-temporal noise reduction, selection of specific informative components (quality enchancement, frequency selection, selection of moving objects). The simplest approaches for dynamic image filtration are purely spatial or temporal

filters. However, image parameters can greatly vary both in time and in space. This require non-stacionary filtration procedures. One of the most effective approaches is adaptive temporal and spatial filtration which shows enchanced quality of image processing (adaptive noise suppression, space-variable temporal filters, rank filters).

Since dynamic image is a three-dimensional signal, it is interesting to develop filters which act simultaneously on space and time. For example, the simplest directed derivative applied to spatio-temporal slices of dynamic image can select stable and moving structures and even produce the object selection by speed. Three-dimensional adaptive filters were used for noise suppression and patterns selection. They act on points lying within spatio-temporal correlation region and appeared to be very stable to moving objects within frame sequence. This can be easily explained: since averaging is performed within adaptively defined spatio-temporal region, the shape of this region is confined along the moving edges in image sequence. Spatio-temporal frequency filters were used to select objects with specific spatial and temporal scales. For example, we have selected short-scale blood flow waves of specific period propagating on skin surface.

5. Functional images. Studies of dynamic images are often concerned with determination of functional structure of dynamic processes. This functional structure can be characterized by spatial distribution of dynamic functions. For example, spatial distribution of oscillations amplitude or spatial regions with homogenuous dynamics are important and can often characterize specific properties of dynamic process. Spatial distributions of dynamic functions are ususally called functional images (it means that they are reflecting functioning of dynamic system rather than field distributions). There are several different ways of constructing functional images.

Distributions of parameters of dynamic processes are the simplest functional images. The general form of construction of this images is the determination of some functional F acting on dynamic image f(x,y,t) and giving some spatial distribution B(x,y)=F{f(x,y,t)}. Parameters used are amplitudes, temporal characteristics (e.g. time constants), speed, spectral power and others. For example, radioisotope studies are concerned mainly with distributions of amplitude and time constants, process of wave propagations can be described by maps of time in which wave front reaches specific points, mapping of electrical brain activity can be characterized by spectral density distributions etc.

More complex functional images are obtained by selecting spatial regions with synchronous temporal behaviour. There can exist both continuous regions and selected sets of points with synchronous dynamics. Such regions are ususally controlled by homogenuous mechanisms and are important for characterization of dynamic processes. If we have selected a set of patterns from each temporal curve then we can introduce some measure of distance between different points of space. Synchronous region is region of space for which this distance, averaged over all points belonging to this region, is less than certain threshold. To find such regions it is usually necessary to use iterative clasterization. However, practical approach is to use apriori information about the image: select "typical" laws of dynamics and "typical" regions of space or extract information about typical temporal dynamics from dynamic gistograms and enchance clasterization starting from this selected regions.

A useful and effective algorithm for construction of functional images is based on dynamic segmentation. Idea of this approach is very simple: use dynamic gistogram H(J,t) and select time dependant thresholds for image segmentation. Segment each image in sequence and obtain spatio-temporal borders between image regions. As the last step find all intersections of this borders in time and obtain spatial image segmentation. It is simple to show that this segmentation is in fact a segmentation into regions with synchronous dynamics. It is necessary to stress that this technique can be used with dynamic images of high dimension and can be applied to select synchronous regions of any type of behaviour (brightness, texture etc).

Knowledge-Based functional images. Recognition of dynamic patterns can be based on specific pattern selection. Usually temporal course of events can be characterized by some qualitative description which is based on knowledge of experimental setup and the sequence of events. For example, it is often important to find small peaks or small falls of signal in selected time intervals ignoring large signal variations during other intervals of time. This qualitative changes can be assigned logical values such as "small decrease", "large rise" etc. Temporal curve then can be described as a sequence of logical clauses and recognition is based on extraction of corresponding sequences in different regions of space and marking spatial regions which satisfy a given set of logical conditions. As every qualitative knowledge-based technique this method of functional imaging appeared to be very effective and gives excellent results selecting the structurally stable patterns from dynamic images.

6.Dynamic structure analysis. Image description in terms of brightness is often inadequate. For a large class of dynamic fields more adequate is study of texture changes in time (remote sensing, surface studies, turbulence, chemistry etc). Two basic approaches were developed. First approach is finding an operator transforming image f(x,y,t) into dynamic vector of patterns Z(p1,p2,...,pN) (texture, size, shape coefficients). Then, study dynamic gistograms of patterns H(p1,t),...,H(pN,t). Such dynamic gistograms quantitatively describe dynamics of structures. The second approach is a little different. As the first step it includes transformation of an image f(x,y,t) to some dynamic parameter field P(x,y,t). For example, field P(x,y) can describe average size of particles in the specific region of space. By investigating dynamics of this field we can use all algorithms of dynamic image processing, for example plot functional images or select regions in which size of particles changes according to some law. It openes vast possibilities of investigation of dynamic spatial structures.

Dynamic image analysis is a new and interesting application of image processing. Future development of this area of research depends upon new applications.

References

1. Image sequence analysis /ed. by T.S.Huang -Springer Verlag, 1982

2. A.M. Taratorin Digital processing of dynamic fields (in Russian) - in Progress in digital optics, Nauka, 1989, Moscow

3. Digital processing of biomedical dynamical images. - in: Advances in biomedical measurements /ed. by E.R.Carson, Pergamon Press, 1988

A LOCAL FILTER FOR IMAGE RESTORATION

H.Jahn , R.Reulke [1]

I. Introduction

The removal of blur from image data is a task of image preprocessing which has been investigated with great effort in the past. Surveys of the problem are given in /1/ and /2/. Because image restoration is an ill-posed problem a small amount of noise can lead to large reconstruction errors, and one needs prior information about the solution to overcome this difficulty. The prior knowlegde of the original scene may be formulated in different manner. In dependence on the prior information and the used criterion of restoration quality one obtains linear or nonlinear restoration algorithms. Of course, nonlinear algorithms may be more powerful then linear ones, but in general the numerical effort is much larger. Often iterative procedures must be used (e.g./3/).

In this paper we obtain a simple linear approximation. The algorithm is a local one. This has the advantage that local changes of the point spread function (PSF) and the scene statistics may be accounted for, which is difficult to do with global algorithms such as the Wiener filter. Further there are also no problems with the the unknown border region and with the accumulation of errors as in iterative procedures. The algorithm to be described in detail here is the linear restoring finite impulse response (F.I.R.) filter, already considered in the past by B.R.Frieden/2/, M.J.Lahart/4/, R.Srinivasan/5/ and others. As in /4/ we use the least mean square solution (LMS) method for the determination of the filter coefficients, but with the difference that subpixel accuracy is considered.

The investigations in this paper are concentrated on the one-dimensional case which is justified for separable PSF's, essentially 1D-PSF's (e.g. image motion blur) and stripe-shaped images. Simulated and real data are used for the demonstration of the results.

II. The FIR restoration filter

We consider the 1D-problem

$$y(x)= \int dx' H(x-x') \cdot \zeta((x') + \xi(x) \tag{1}$$

where x is the spatial coordinate and $H(x)$ is the space invariant impulse response or point spread function (PSF) of the whole system or of system parts only. The measurement values $y(i\Delta)$ are the sampled values of $y(x)$ at the positions $x_i=i\cdot\Delta$. $\zeta(x')$ is the original scene to be restored and $\xi(i\cdot\Delta)$ are

[1] Institute of Space Research
Rudower Chaussee 5, Berlin 1199, GDR

the measurements errors (noise) which we considerd to be uncorrelated random numbers with vanishing expection values:

$$\langle \xi(i\cdot\Delta)\rangle=0 \quad , \quad \langle \xi(i\cdot\Delta)\cdot\xi(j\cdot\Delta)\rangle=\sigma^2\delta_{i,j} \tag{2}$$

The function $\zeta(x)$ is considered as a stationary random process with certain first and second order moments which yields the prior information about the scene. This information is only necessary for the stabilization of the ill-posed problem and therefore it must not be known precisely. It is better to use coarse prior information if one has no exact knowledge. If the statistics of the scene and the noise is known and the measurement values $y(i\Delta)$ are given then one can estimate each value $\zeta(x)$ using a linear combination of some of the values $y(i\cdot\Delta)$. From the very beginning we take into account only a finite number of measurement values to lower the calculation effort. So, for the estimation of the value $\zeta((i+\varepsilon)\cdot\Delta)$ $\left(-\frac{1}{2} \le \varepsilon \le \frac{1}{2}\right)$ we choose the linear FIR filter

$$\zeta((i+\varepsilon)\Delta) = a(\varepsilon) + \sum_{l=-N_o}^{N_o} b_l(\varepsilon)\cdot y((i+l)\Delta) \tag{3}$$

The unknown coefficients a, b_l can be determined by the minimization of the least mean square error

$$\langle(\zeta((i+\varepsilon)\cdot\Delta) - \hat{\zeta}((i+\varepsilon)\cdot\Delta))^2\rangle.$$

Although for this procedure a discretization of the integral (1) is not necessary we do it for the sake of further simplification. Choosing the step-wise approximation $\zeta(x)=\zeta(j\cdot\delta)$ for $j\cdot\delta \le x \le (j+1)\delta$ we get from (1) the approximation

$$y(i\Delta) = \sum_j h(i\Delta-j\delta)\cdot\zeta(j\delta) + \xi(i\Delta) \tag{4}$$

with

$$\int_{j\delta}^{(j+1)\delta} dx'\, H(i\Delta-x') = \int_0^{\delta} dx\, H(i\Delta-j\delta-x) = h(i\Delta-j\delta) \tag{5}$$

Choosing $\delta < \Delta$ we have subpixel accuracy. In this case the discretized PSF $h(i\Delta-j\delta)$ is not translation invariant. For abbrevation we write

$$\Delta = \nu\cdot\delta \tag{6}$$

with ν a positive integer. $\nu=1$ means pixel accuracy, which is the most crude approximation, leading to translation invariance. For the estimation of the value $\zeta(i\Delta-m\delta) = \zeta((i\nu-m)\delta)$ in accordance with (3) we choose the linear FIR filter

$$\hat{\zeta}(i\nu-m) = a(m) + \sum_{l=-N_o}^{N_o} b_l(m)\cdot y((i+l)\nu) \tag{7}$$

Here for simplicity the subpixel resolution parameter δ has been omitted. Choosing ν as an odd integer, we calculate $\hat{\zeta}$ at the subpixel position

$m=0, \pm1, \ldots, \pm\frac{\nu-1}{2}$

for each i (see figure 1 for $\nu=3$)

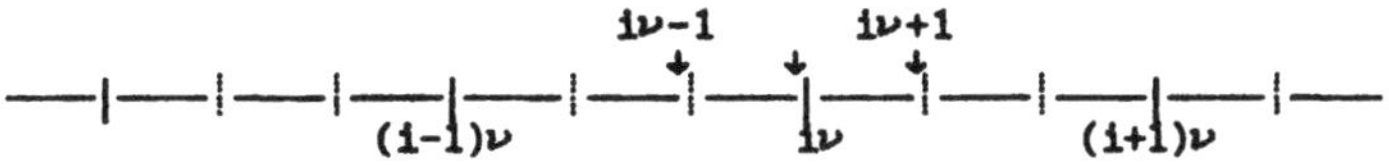

Figure 1

Because for $\nu \neq 1$ the problem is not translation invariant, the coefficients a, b_k of the filter (7) depend on the subpixel position m. Therefore for each m a separate filter is needed.

For the determination of the filter coefficients the mean square error

$$\Im(a, b_{-N_o} \ldots b_{N_o}) = \langle [\hat{\zeta}(i\nu - m) - \zeta(i\nu - m)]^2 \rangle \tag{8}$$

has to be calculated.

Supposing

$$\sum h(k) = 1 \tag{9}$$

and writing

$$\zeta(j) = \mu + \Delta\zeta(j) \quad ; \quad \mu = \langle\zeta(j)\rangle \tag{10}$$

we get from (7) and (4)

$$\hat{\zeta}(i\nu - m) = a + \mu \cdot \sum_l b_l + \sum_l b_l \cdot \sum_k h(k) \cdot \Delta\zeta((i+1)\nu - k) + \sum_l b_l \cdot \xi((i+1)\nu)$$

Inserting this expression into (8), introducing the autocovariance matrix

$$\langle \Delta\zeta(j) \cdot \Delta\zeta(j') \rangle = \varkappa^2 \cdot \gamma(j - j') \quad ; \quad \gamma(0) = 1 \tag{11}$$

which describes the smoothness of the scene, and using (2) the result is

$$\Im = a^2 + 2\mu a(\sum b_l - 1) + \mu^2(\sum b_l - 1)^2 + \varkappa^2 + \sigma^2 \sum b_l^2 +$$
$$+ \varkappa^2 \sum_{l,l'} b_l \cdot b_{l'} \cdot \sum_{k,k'} h(k) \cdot h(k') \cdot \gamma((l-l')\nu - (k-k')) - 2\varkappa^2 \sum_l b_l \sum_k h(k) \cdot \gamma(l\nu - (k-m)) \tag{12}$$

The condition $\partial\Im / \partial a = 0$ leads to $a = \mu \cdot (1 - \sum_l b_l)$ (13)

Using this, from $\partial\Im / \partial b_n = 0$ we get

$$\sum_{l=-N_o}^{N_o} \left\{ \sum_{k,k'} h(k) \cdot h(k') \cdot \gamma((l-n)\nu + k' - k)) + \frac{\sigma^2}{\varkappa^2} \cdot \delta_{n,l} \right\} \cdot b_l = \sum_k h(k) \cdot \gamma(n\nu - k + m) \tag{14}$$
$$(n = 0, \pm 1, \ldots \pm N_o)$$

This is a set of linear equation which must be solved to obtain the optimal filter coefficients $b_l = \hat{b}_l(m)$.

The parameter $\varepsilon = \sigma^2 / \varkappa^2$ which influences the main diagonal elements of the matrix

$$A_{n,l} = \sum_{k,k'} h(k) \cdot h(k') \cdot \gamma((l-n)\nu + k' - k) + \varepsilon \cdot \delta_{n,l} \tag{15}$$

is essential for the regularization of the solution. If one has no knowlege about the dispersions σ and $\varkappa$, ε may be chosen experimentally in order to obtain a good solution. In most cases the autocovariance matrix γ is not known too. Then one can make the special choice

$$\gamma(j) = \delta_{j,o} \tag{16}$$

which allows for the fastes change of the solution. In this case instead (14) and (15) we obtain

$$\sum_{l=-N_o}^{N_o} A_{n,l} \cdot b_l = h(n\nu + m) \quad \left[n = 0, \pm 1, \ldots, \pm N_o \; ; \; m = 0, \pm 1, \ldots \pm \frac{\nu - 1}{2} \right] \tag{17}$$

with

$$A_{n,l} = \sum_k h(k) \cdot h(k + (n-l)\nu) + \varepsilon \cdot \delta_{n,l} \tag{18}$$

It should be said that the matrix **A** is only a function of n-l which alleviates its computation. Further $\mathbf{A}$ does not depend on the subpixel position m. Therefore $\mathbf{A}$ must be inverted only once for all filters $m=0,\pm 1,..,\pm \frac{\nu-1}{2}$. The filter coefficients a, b_l may be calculated in advance for a given imaging system and a certain class of scenes. Once the coefficients are given one only needs $2N_0+1$ multiplications and additions for the calculation of one subpixel scene value $\hat{\zeta}(i\nu-m)$.

The derived filter has an essential disadvantage: If the scene data are constant values, i.e. $\zeta(j\delta) = \zeta_0$ = const. then according to (4) from (7) and (13) we get in the absence of noise

$$\hat{\zeta}(i\nu-m) = \mu\cdot\left[1-\Sigma\hat{b}_l(m)\right] + y_0\cdot\sum_{l=-N_0}^{N_0}\hat{b}_l(m)$$

Only for $\Sigma\,\hat{b}_l(m)=1$ the true result $\hat{\zeta}(i\nu-m) = y_0$ is obtained. Because we didn t derive an exact solution of the problem the sum $\sum_l \hat{b}_l$ is not exactly equal to one. To obtain an exact reconstruction of constant grey values, we normalize the $\hat{b}_l'$. Then according to (14) the constant value $\hat{a}$ vanishes and we obtain the filter

$$\hat{\zeta}(i\nu-m) = \sum_{l=-N_0}^{N_0}\hat{b}_l(m)\cdot y((i+l)\nu)\ . \tag{20}$$

Of course this procedure slightly enhances the mean square error if the assumptions about the scene and noise statistics are fulfilled, but in general one obtains better results.

In figure 2 the filter coefficients $\hat{b}_l(0)$ together with the Gaussian PSF as the solution of (17), (18) are shown for $N_0=10, \varepsilon=0.01, \sigma_{PSF}=2$.

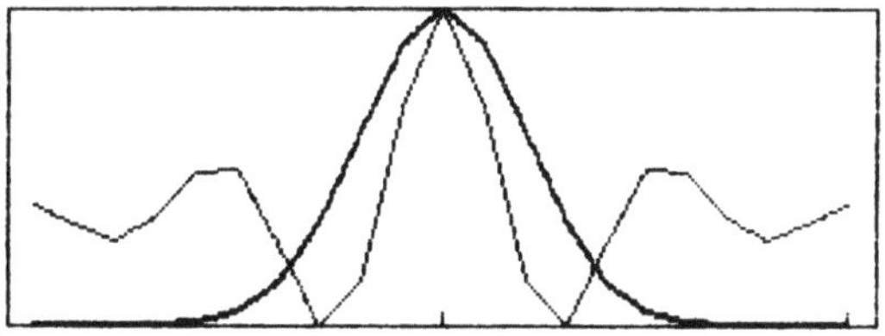

Figure 2

So, using the method of least mean square error, we have obtained a simple linear FIR restoration filter which is easy to implement on conventional image processing systems. For Gaussian random processes ζ and ξ it is the best one as compared with all other filters, including nonlinear ones, which use only 2N+1 samples $y(j\Delta)$ too. If the underlying processes are not Gaussian, which in most cases is true for the scene data $\zeta(j)$, then the derived filter is the best linear one.

IV. **Results**

First we consider the linear FIR restoration Filter for $\nu=1$ (pixel accuracy). As already mentioned above it is an optimal filter for Gaussian processes ζ and ξ. Therefore it is useful first to investigate the behavior of the filter for such processes. The statistically independent noise and scene variables

$\xi(i\Delta)$, $\zeta(j\Delta)$ are generated by a Gaussian pseudorandom number generator. According to (4) the "measured" values are obtained by

$$y(i) = \sum_{j=i-M_o}^{i+M_o} h(i-j)\cdot\zeta(j) + \xi(i), \tag{20}$$

where M_o is an integer, which describes the truncation of the PSF **h** at small values.

In figure 3 and 4 the noise to signal ratio was 5% and the PSF was a Gaussian one.

Figure 3a show the process $\zeta(i)$ together with the values $y(i)$ (M_o=3,σ_{PSF}=1).

The restauration filter with ε=0.01 and N_o=3 leads to the result of figure 3b which shows a considerable improvement of the data.

3a

3b

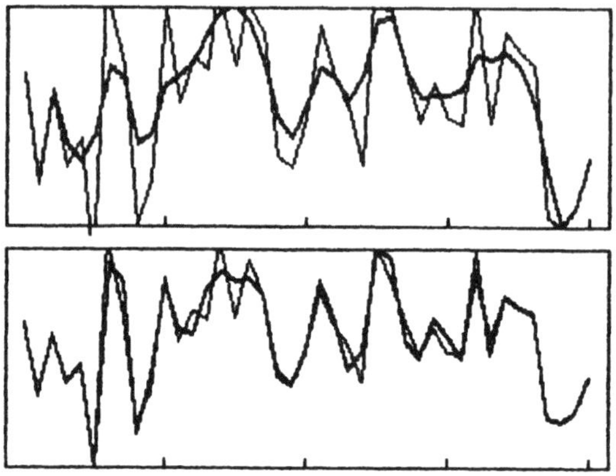

Figure 3

Now we consider more realistic signals $\zeta(i)$ which are generated by the recursion

$$\zeta(i+1) = \zeta(i) + \eta(i) \;, \tag{21}$$

where the $\eta(i)$ are uncorrelated random variables with the Non-Gaussian probability density

$$p(\eta) = P_o\delta(\eta) + (1-P_o)\cdot\frac{1}{\varkappa_o\sqrt{2\pi}}\cdot e^{-\frac{\eta^2}{2\varkappa_o^2}} \tag{22}$$

According to (22) with the probability P_o the value $\eta(i)=0$ is generated at each step and therefore we have $\zeta(i+1) = \zeta(i)$ in most cases, if P_o is near unity. Because with the probability $1-P_o$ a value $\eta(i)\neq 0$, which obeys a Gaussian distribution with the dispersion $\varkappa_o$, is generated, jumps of different heights arise at certain positions i. Therefore, the Non-Gaussian Markov processes, generated by the recursion (21) are suitable models for a wide-spread class of scenes.

Fig 4a represents such a process for P_o=0.96, $\varkappa_o$=20 together with the Gaussian-blurred data (σ_{PSF}=2, M_o=10). The output of the restoration filter with ε=0.01, N_o=10 is shown in figure 4b. One can observe a considerable sharpening of the edge but also anenhancement of noise.

Now a real data set, which has been obtained by a comercial infrared imaging camera, is considered.

4a

4b

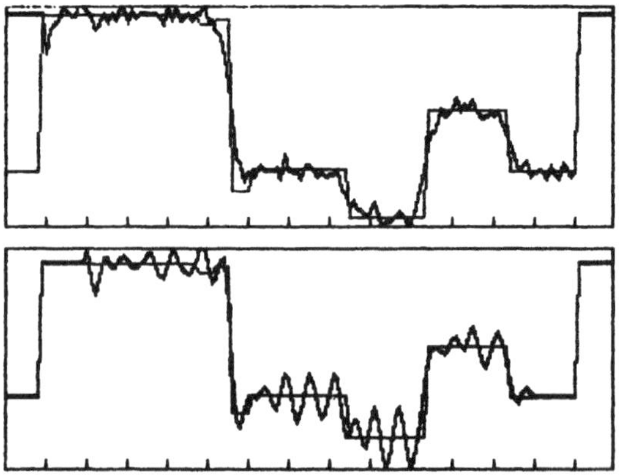

Figure 4

Measurements for the determinatin of the PSF and the noise of the IR-system have been done.

Figure 5 shows a line scan of a measured edge (dashed line). To demonstrate the restoration capability of the algorithm the optics of the camera was defocused. The PSF of the IR-system was modelled as a trapeze with a spread over 15 pixel in the image plane. The regularization parameter was $\varepsilon=0.01$, which is roughly a noise-signal-ratio of 5%. Application of the described algorithm on the measured edge yields the full line. The edge is now sharpened and the noise is greater as before especially in the edge region.

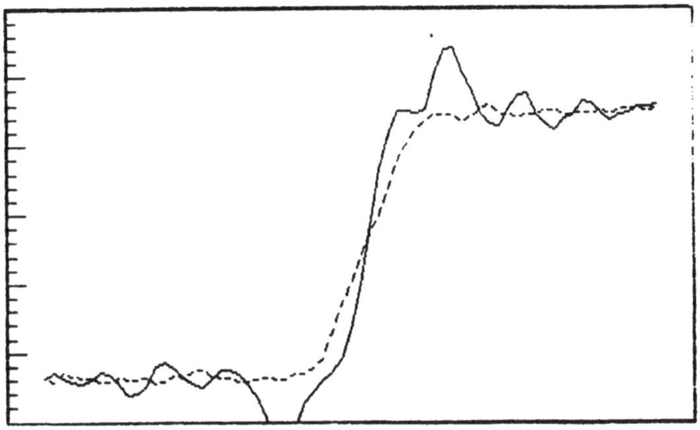

Figure 5

Now subpixel accuracy is considered. The data of **fig.6** have been obtained by the convolution of the signal shown in fig.6a with a Gaussian PSF ($\sigma_{PSF}=5, M_O=10$) and the addition of 1% noise ($\sigma=0.01$). The output of the linear restoration filter with $N_O=3$, $\varepsilon=10^{-3}$, $\nu=5$ shown in fig.6b represents the original signal essentially better then the blurred and noisy data.

6a

6b

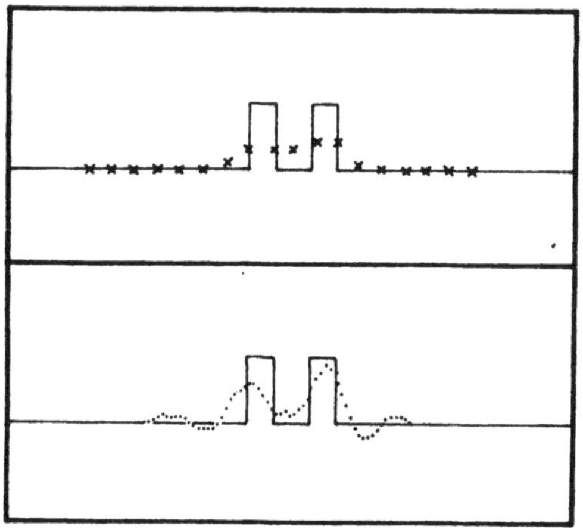

Figure 6

References

/1/ H.C.Andrews, B.R.Hunt; Digital Image Restoration Prentice-Hall, Inc., Englewood Cliffs, New Jersey 1977

/2/ T.S.Huang(Ed.); Picture Processing and Digital Filtering; Topics in Applied Physics, Vol.6; Springer-Verlag 1979

/3/ R.W.Schafer, R.M.Mersereau, M.A.Richards; Constrained Iterative Restoration Algorithm Proceedings of the IEEE, Vol.69, 432-450(1981)

/4/ M.J.Lahart; Local image restoration by a least squares method; J.Opt.Soc.Am.69, 1333-1339(1979)

/5/ R.Srinivasan; Image restauration by spatial filter design Proc.of SPIE, Vol.707; 193(1986)

TOMOGRAPHIC APPROACH TO IMAGE LINEAR BOUNDARY DETECTION

I.G. Kazantsev, V.P. Pyatkin[1)]

1. INTRODUCTION

An important problem in image processing and scene analysis is the detection of edges used to identify objects, interpret contours, analyze texture and solve other computer vision tasks. Most of edge detection techniques implemented in vision systems are based on the digital approximations of gradient's type expressions, such as Robert's, Sobel's, Prewitt's and other [4] operators. It is well known that differentiation of discretized function is an ill-conditioned problem in Hadamard sense and this mathematical operation is very sensitive to noise.

Therefore, investigations of some variants of gradient type operators which are more robust continue up to now. An obvious way is to smooth an image before taking derivatives; the best known successfull example of such a combination of presmoothing and differentiating is the Marr-Hildreth operator [3]. Edge detection by differential analysis of the image always leads to the classical problem of compromise between noise reduction and resolution. The simplest parameter of smoothing usually is the size of scanning window.

In contrast to image differential analysis there exists another group of methods using in large part integrals: Hough transform [1] and Hueckel operator [2]. The method we suggest in this paper belongs to this group and has similar mathematical properties. We start our discussion by considering the informative projection choice criterion in computed tomography which for edge detection problem serves as an indicator of image homogeneity direction.

2. RADON TRANSFORM AND INFORMATIVE PROJECTION CHOICE CRITERION

Let f be a square integrable function which vanishes outside of the unit disk D on the plane R^2, i.e. $f \in L_2(D)$. For an arbitrary $\alpha \in [0,2]$ we define rotation of the function f as a function f_α:

$$f_\alpha(x,y) = f(x \cdot \cos\alpha + y \cdot \sin\alpha,\ -x \cdot \sin\alpha + y \cdot \cos\alpha).$$

Integral of the function f_α taken along the line $\{(x,y) : y \in R,\ x = t\}$ which corresponds to the chosen value of parameter $t \in [-1,1]$ is called projection of the function f_α. Considering projections for each angle α as a function $p_\alpha(t)$ yields the Radon transformation of the function f:

$$P_\alpha f(t) = p_\alpha(t) = p(\alpha,t) = \int f_\alpha(t,y)dy.$$

The objective of a discrete reconstruction problem is to determine an approximation of the unknown function $f(x,y)$ given a finite set of the projections $P_{\alpha_1} f,\ P_{\alpha_2} f, \ldots,\ P_{\alpha_m}$.

---

1) Computing Center, Siberian Division, USSR Academy of Sciences, Academician Lavrentiev, 6, 630090, Novosibirsk, USSR.

The projection maximizing the functional

$$Q(P_\alpha) = Q(f, \alpha) = 1/2 \int_{-1}^{1} (1 - t^2)^{-1/2} P_\alpha^2(t)dt$$

gives the largest contribution to this reconstruction.[1]

We can assert that in general $Q(f, \alpha) \leq \|f\|^2$; if there exists an α for which $f_\alpha(x,y) \equiv g(x)$ on D for some g (i.e., the function f_α can beregarded as a function of single variable), then $Q(f, \alpha) = \|f\|^2$. On the contrary, if there is no such α, then $Q(f, \alpha) < \|f\|^2$. Consequently, if the norm $\|f\|$ is known, the functional Q may serve as a criterion of homogeneity of the function f in the corresponding direction $(\alpha + \pi/2)$.

We call the projection p_α which maximizes the criterion Q 'the most informative projection' (in the sense of L_2 norm).

Numerical computation of informative projection choice criterion Q may be implemented with the Gauss-Tchebyshev formula:

$$\int_{-1}^{1} \frac{u(x)dx}{\sqrt{1 - x^2}} \approx \sum_{k=1}^{n} a_k u(x_k),$$

where $a_k = \pi/n$; $x_k = \cos(\pi(2k - 1)/n)$, $k = 1,\ldots,n$. Note that in this approximation we avoid arithmetic operation of division.

3. STRAIGHT EDGE DETECTOR

The basic idea of suggested straight edge detector is to introduce such a model of edge which considers direction as a main sign of the edge and computes it. For this purpose let us consider sentences

A(f) = "An image f(x,y) does not depend on variable y",

B(f) = "For $\alpha = 0$ the equality $Q(f, \alpha) = \|f\|^2$ holds";

as we have seen, they are identical and express the same property of homogeneity of the image f in the direction of axis y. Since image usually considered as an edge paradigm (e.g. step function, roof function and others) satisfy these two predicates we introduce the general concept of the straight line edge in an image by defining a set

$$F = \{f \in L_2(D) : A(f) \text{ is true}\}$$

So, the first sentence serves as a definition of generalized lines, or lineaments, and the second one may be searched as a criterion of the finding lineament direction. We choose an angle $\alpha + \pi/2$ as a possible homogeneity direction of the searched function f equality $Q(f, \alpha) = \|f\|^2$ holds.

The advantage of using such a model is that convolution of $f \in F$ with any function h (for which $f*h$ exists) belongs to F, hence the prob-

---

1) Kazantsev, I.: Radon Space-Based Straight Edge Detection in Digital Images. Comp.and Art.Intel. 2 (1989), (be printed).

lem of edge blurring arises in the following edge direction finding stages. However for image f with the property $f(x,y) = \text{const}$ we have $Q(f, \alpha) = \|f\|^2$ for arbitrary α; moreover, noisy effects make direction search procedure nonstable. This gives an impetus to a general analysis of the informative projection choice criterion.

The following easy to verify corollaries are useful for the new straight edge detector constructing.

Corollary 1. If $f \in F$ and $\alpha \in [0, \pi/2)$ then

$$p(\alpha, x) = p(-\alpha, x);$$
$$p(\alpha + \pi/2, x) = p(\alpha - \pi/2, x).$$

Corollary 2. If $f \in F$ and $\alpha \in [0, \pi/2)$ then

$$Q(f, \alpha) = Q(f, \alpha + \pi);$$
$$Q(f, \pi/2 - \alpha) = Q(f, \pi/2 + \alpha).$$

Corollary 3. If $f, g \in F$ and f is odd, g is even, then

$$Q(f + g, \alpha) = Q(f, \alpha) + Q(g, \alpha).$$

Corollary 4. If $f(x,y) = \sin(kx)$ then

$$Q(f, \alpha) = (1 + J_0(2k) - J_0(2k\sin\alpha) - J_0(2k\cos\alpha))/2k^2\sin^2\alpha;$$

If $f(x,y) = \cos(kx)$ then

$$Q(f, \alpha) = (1 - J_0(2k) - J_0(2k\sin\alpha) + J_0(2k\cos\alpha))/2k^2\cos^2\alpha;$$

where J_0 is Bessel's function of the first kind.

Corollary 5. Value $\alpha = 0$ maximizes the difference

$$\Delta(f, \alpha) = Q(f, \alpha) - Q(f, \alpha + \pi/2)$$

for an arbitrary function $f \in F$.

These corollaries explain us some features of the criterion Q behavior in the case when generalized lines from F are investigated. Let us denote some images which seem as lines for human visual system:

1. Zigzag -  $g_1(x,y) = \begin{cases} x+1, & x \leq 0 \\ x, & x > 0 \end{cases}$

2. Fin -  $g_2(x,y) = \begin{cases} 0, & x \leq 0 \\ 1-x, & x > 0 \end{cases}$

3. Roof -  $g_3(x,y) = \begin{cases} 1+x, & x \leq 0 \\ 1-x, & x > 0 \end{cases}$

4. Step -  $g_4(x,y) = \begin{cases} 1, & x \leq 0 \\ -1, & x > 0 \end{cases}$

5. Ramp -  $g_5(x,y) = \begin{cases} x+1, & x \leq 0 \\ 1, & x > 0 \end{cases}$

It is also interesting to investigate images which don't look like lines but belong to the set F:

6. Sinus -  $g_6(x,y) = \sin(x)$

7. Slope -  $g_7(x,y) = x$

8. Plateau- $g_8(x,y) = \text{const}$

Corollary 6. For images g_1-g_5 informative projection choice criterion Q is nonincreasing function of α on interval $[0, \pi/2]$; values $\alpha = 0$ and $\alpha = \pi/2$ maximize and minimize the criterion correspondingly.

So, bearing on the analysis of criterion features we can suggest a number of lineament direction indicators:

$$\min_{\alpha} (\|f\|^2 - Q(f, \alpha)), \quad \max_{\alpha} (Q(f, \alpha) - Q(f, \alpha + \pi/2)), \ldots$$

All these detectors have a disadvantage that, generally, image f may have projections which are identically zero and for this reason there will be a number of directions which give an extremum for detectors. Hence, lineament detector E must contain itself information about criterion Q in a general, integral form and meet the following requirements:

1. $0 \leqslant E(f) \leqslant 1, \quad f \in L_2(D)$;
2. $E(f + \text{const}) = E(f)$;
3. $E(f \cdot \text{const}) = E(f)$;
4. $E(\text{const}) = 0$;
5. There exists a threshold T such that image f for which $E(f) \leqslant T$ does not contain straight edge from the viewpoint of human perception.

As a result, we suggest new straight edge detector of the form:

$$E(f) = 1 - \left(\int_0^{\pi} Q(f, \alpha) d\alpha - \min_{\alpha} Q(f, \alpha)) / \pi (\max_{\alpha} Q(f, \alpha) - \min_{\alpha} Q(f, \alpha) \right)$$

The detector E has values for the images g_1-g_8:

$E(g_1) = 0.80$	$E(g_5) = 0.53$
$E(g_2) = 0.73$	$E(g_6) = 0.49$
$E(g_3) = 0.67$	$E(g_7) = 0.47$
$E(g_4) = 0.58$	$E(g_8) = 0.00$

We assume that edge threshold T is approximately 0.50; detector E introduces an order relation into the set F: if an image f looks like straight border strongly than image g, then $E(f) > E(g)$.

So, the technique of edge direction detection is in computing the criterions Q and E, then if $E(f) > 0.50$ we deduce that f presents an edge with direction $\alpha = \arg\max_{\theta} Q(f, \theta) + \pi/2$.

4. COMPUTER IMPLEMENTATION AND HUECKEL OPERATOR

The digital image $\varphi(x,y)$ of interest to us is scanned by the window $D = \{s,t: (x - s)^2 + (y - t)^2 \leqslant r^2\}$ and the fragment $f \in L_2(D)$ such that $f(s,t) = \varphi(s,t)$ for $(s,t) \in D$ is searched.

For every point (x,y) of the image φ values

$$\alpha(x,y) = \arg\max_{\theta}(Q(f,\theta)),$$

$$\mathcal{E}(x,y) = E(f)$$

are computed and stored. Magnitudes $\mathcal{E}$ and α are amplitude and phase characteristics of an edge in the fragment f. When scanning is over, threshold T is chosen interactively and if $\mathcal{E}(x,y) \geqslant T$ then straight line of direction $\pi/2 + \alpha(x,y)$ and 2r in length is visualized, where (x,y) is the central point of the line.

Let us look at how suggested approach and Hueckel edge detector are related. The Hueckel operator is a well known technique for the detection of lines in digital images also using a circular window and an integration as a basic computational procedure. Hueckel used an ideal edge which is characterized by the step function

$$G(x,y,\alpha,p,a,b) = \begin{cases} a, & x\cdot\cos(\alpha) + y\cdot\sin(\alpha) \leqslant p \\ b, & x\cdot\cos(\alpha) + y\cdot\sin(\alpha) > p \end{cases}$$

to approximate best a given empirical image f from Hilbert space $L_2(D)$. Note, that it is possible to eliminate the parameters a and b because a visual system is interested not in absolute values of edge amplitude but in the knowledge: whether a border exists in a given point of the image. So we use the Hueckel edge G of the form $G = g_4(x,y)$ and vary only parameter α:

$$\|f_\alpha(x,y) - g_4(x,y)\|^2 = \|f\|^2 + \|g_4\|^2 - 2\int_{-1}^{1} p(\alpha,x)g_4(x,y)dx$$

It follows that for the Hueckel edge finding we can use the simpler form of criterion Q:

$$Q(f,\alpha) = 2\left[\int_{-1}^{0} p(\alpha,x)dx - \int_{0}^{1} p(\alpha,x)dx\right] - \|g_4\|^2.$$

Consequently, other types of edges are ignored but computational simplicity and prevalence of the step edges make this fast version of the Radon space-based straight edge detector efficient and useful.

Computer implementation problems of proposed approach are considered; the results of this approach application to the images of different nature are demonstrated.

References

[1] Duda, R., P. Hart: Use of the Hough Transform to Detect Lines and Curves in Pictures. Comm.ACM 15 (1972), 11-15.
[2] Hueckel, M.: An operator which locates edges in digitized pictures. J.ACM 18 (1971), 113-125.
[3] Marr, D., E. Hildreth: Theory of Edge Detection. Proc.R.Soc.Lond.B. 1980, 187-217.
[4] Pratt, W.: Digital Image Processing. J.Wiley, New York 1978.

A THEORETICAL BASIS FOR THINNING METHODS

Ulrich Eckhardt

1. Introduction

One of the most widely used methods for preprocessing of binary images is thinning. The popularity of this method rests on the fact that a considerable data reduction is achieved while retaining "essential" properties of the original image. Moreover, topological features, which cannot be verified by a genuinely parallel method, are more easily treated in thinned images |10|. For these reasons, many articles on this topic were published. Most of them are concerned with modifications of existing methods in order to yield "nicer" results out of the thinning process. Also, many results of numerical experiments are available in the literature. In contrast, there exist only few theoretical investigations concerning the foundations for the different methods or explaining their properties (see |12|, |13|, |15|). Since thinning is a fundamental process in digital picture processing, parallel implementation becomes mandatory. Almost all of the existing methods for parallel thinning, however, use only simple ad-hoc methods for parallelization.

The aim of this report is first to state some requirements which are considered to be essential for thinning methods. Based on these requirements, a classification and evaluation of existing methods can be found. A special class of thinning methods is presented and its theoretical properties are derived.

2. Basic Definitions

The *digital plane* is the set of all points in the plane having integer coordinates. A *digital set* is any finite subset of the digital plane. For any point $P = \binom{i}{j}$ in the digital plane, the *direct (or 4-) neighbors* of P are the points $Q = \binom{k}{l}$ such that $|i - k| + |j - l| = 1$. The *indirect neighbors* of P are points Q with $|i - k| = |j - l| = 1$. All neighbors together are also called *8-neighbors*.

By means of the neighbor-relation, a topology is defined in the digital plane |13|. As usual we assume the foreground to be equipped with the 8-topology, i.e., two points in a digital set S are connected if there exists a sequence of points in S starting at one of the given points and ending at the other such that each two successive points in the sequence are 8-neighbors of each other. The complement of S is analogously equipped with the 4-topology. From this definition it becomes clear what is meant by *connection components* of S or of its complement.

For any point P in a digital set S the *connection number* $C_S(P)$ of P with

Institut für Angewandte Mathematik der Universität Hamburg,
Bundesstraße 55, D-2000 Hamburg 13

respect to S is the number of 8-components in the intersection of S with the set of all neighbors of P. A *simple point* is a point in S having connection number one and at least one direct neighbor not in S. An *end point* is a point in S having exactly one neighbor in S. A *strict boundary point* P is a simple point with the property that the intersection of S with the set of all neighbors of P is even 4-connected.

3. Thinning Methods

A (sequential or parallel) *thinning operator* is a mapping T associating to each digital set S a digital set T(S) such that for all finite digital sets S

(T1) T(S) is contained in S,

(T2) T is *correct*, i.e., S and T(S) have the same number of (8-) connection components and the complements of S and of T(S) have the same number of (4-) connection components.

For any finite digital set S there exists a number n such that $T^n(S) = T^{n+1}(S)$. The set $T^n(S)$ is called the *skeleton* of S.

A thinning operator T is correct if $P \in S - T(S)$ implies that P is a simple point of S. Usually, therefore, it is required that all points of S - T(S) are simple points or even strict boundary points or belong to a subset thereof. The following Theorem is easily proved (|5|, Satz 1.1):

Theorem 1: For $P \in S \cap T(S)$ is $C_S(P) \leq C_{T(S)}(P)$ whenever each end point Q which is an 8-neighbor of P also belongs to T(S).

Theorem 1 is the theoretical justification for excluding end points from elimination since this implies that any point with $C_S(P) > 1$ belongs to the skeleton of S. We therefore require in addition for a thinning operator

(T3) If $P \in S - T(S)$ then P is simple but no end point.

Any thinning operator T fulfilling condition (T3) can be organized in the following way:

(C1) Topological Criterion: A subset of S, the set of *candidates* for elimination, is identified. This set is contained in the set of all simple points of S and it contains no end points of S. We denote the set of all candidate points of S by can(S): S - can(S) is contained in T(S).

(C2) Collision Avoidance: A *selection rule* is applied to each candidate point deciding whether the latter belongs to T(S) or not. The selection rule is chosen as to guarantee that (T2) is fulfilled.

Examples for selection rules are

- Sequential thinning methods: The rule consists in selecting exactly one of the candidate points.
- Parallel thinning methods:

Local four phase method |14|,
Global four phase method |9|,
Local two phase method (Rutovitz' method, see |3|),
Global two phase method |6|.

An operator R mapping subsets of the digital plane into subsets of the digital plane is said to be *motion invariant*, if R commutes with all translations and rotations mapping the digital plane into itself.

Theorem 2: If the operator $S \to can(S)$ is motion invariant, then the thinning method $T(S) = S - can(S)$ is correct.

(For a proof of this Theorem see |6|).
The assertion of the Theorem implies that in case of a motion invariant rule (C1) for identifying candidates for thinning special precautions for avoiding collisions in parallel implementation (C2) are not necessary.

4. Perfect Points

Many authors proposed to choose as candidates for thinning only simple points having an interior point as 8-neighbor (|11|, |8|, |16|, |7|).
An *interior point* of a digital set S is a point P in S such that all direct neighbors of P are also in S. A point P in S is said to be *perfect*, if it has a neighbor which is an interior point and if all direct neighbors of P which are not neighbors of this interior point, are not in S.

We now investigate the method T_o consisting in the topological criterion (C1): $P \in can(S)$ if and only if P is simple and perfect. Obviously this criterion is invariant with respect to motions, hence Theorem 2 applies. We note some properties of this thinning method:

Any simple and perfect point is a strict boundary point (|2|, Theorem 3.2).

End points are never perfect, hence there is no need to treat them separately and Theorem 1 holds for method T_o.

Configurations of the form (Tamura |15|)

```
x x                          x x
  x x                        x x
    x x                      x x
      x x                    x x
         . .                 : :
          . .                : :
```

contain no perfect points. If these configurations are processed by a usual thinning method (e.g. sequentially from the upper left corner to the lower right corner line by line), they will be completely eroded. Since such structures sometimes contain significant information, this behaviour is not acceptable.

We note, that method T_o, due to its essential parallel nature, is well-defined, i.e., its results do not depend on the specific implementation used. This is not the case for most existing thinning methods.

Moreover, method T_o compares favourably with other thinning methods from a theoretical point of view. Consider the situation that a point is simple but not perfect and thus is not eliminated by T_o but can be a candidate for any other method.

Theorem 3: Let P be a point in a digital set S which is simple but not perfect.

1. If P has at most two direct neighbors in S then P remains simple under thinning rule T_o.
2. If P has three direct neighbors in S then there exists a direct neighbor Q of P in S such that $C_S(Q) < C_{S-\{P\}}(Q)$.

(See |5|, Satz 6.1). In the first case, P remains simple and might be treated at the end of the thinning process by a suitable postprocessing method. In the second case, elimination of P leads to a result which is certainly not motion invariant when Q happens to be simple. Hence Theorem 3 means that thinning rule T_o is in some sense best possible if only information about neighbors of a point is available together with a "status" information (i.e. whether a point is interior point or not). Any deviation from T_o either can be postponed or else it leads to a decision whose result can be not motion invariant and which is therefore arbitrary.

For a more close theoretical investigation of the properties of method T_o it seems appropriate to study interior points of digital sets. The *kernel* of a digital set is the set of all its interior points. A point on the *boundary* (i.e., it is not an interior point, but belongs to the set) is said to be a point on the *inner boundary* if it has an 8-neighbor which is an interior point. There is a close relationship between simple and perfect points and the structure of both boundaries. Any point on the boundary and also on the inner boundary has a uniquely determined successor on the boundary and the inner boundary, respectively. These successors can be found in a straightforward way by means of a boundary-following procedure (see |5|, § 2). Analogously, there exist uniquely determined predecessors. It can be shown that a point on the inner boundary is simple and perfect if its successor on the inner boundary coincides with its successor on the boundary and if the same holds also for the predecessors (|5|, Satz 4.1).

5. Irreducible Sets

Arcelli |1| proved by means of an example that there exist digital sets containing no simple points but having a nonempty kernel. This means that it is not possible to reduce a digital set to a "graph-like" structure by thinning. For understanding the properties of thinning methods it is therefore necessary to study irreducible sets. A digital set is said to be *s-irreducible* ("sequential") if it contains no simple points. It is said to be *T_o-irreducible* if it contains no simple and perfect points.

<u>Theorem 4</u>: The (8-) connection components of the kernel of a (finite) (s- or T_o-) irreducible set are simply connected.
The inner boundary belonging to a component of the kernel is composed of three different types of boundary elements. These are generated by 90° rotations from the following configurations (x = black, o = white, # = interior point):

1. Vertex elements

```
. . x . .          . x x .
o x # x o          x # # x
x o x o x          o x x o
```

2. Diagonal line element

```
         x .·
       x # x
     x # x o
   x # x o x
 x # x o x
.· x o x
```

The inner boundary corresponding to a connection component of the kernel is composed of four diagonal line elements with different directions. Two line elements are joined by a vertex element such that a left turn is executed when following the inner boundary in the direction leaving the kernel on the left-hand side.

For a proof of this Theorem see |4|, Theorem 5.1 (for s-irreducible sets) and |2|, Theorem 4.1 (for T_o-irreducible sets).

A *1-hole* of a digital set is a point not in S whereas all its direct neighbors belong to S. The irreducible sets become very simple if it is required that they do not contain 1-holes. This is the theoretical justification for removing 1-holes before thinning.

<u>Theorem 5</u>: The kernel components of a T_o-irreducible set containing no 1-holes have only the following types (and those generated therefrom by 90° rotations):

1.
```
x
```
2.
```
x x
```
3.
```
  x
x
```
4.
```
  x
x x
```
5.
```
x x
x x
```
6.
```
  x
x x
x
```
7.
```
  x
x x x
  x
```
8.
```
  x
x x x
  x x
```
9.
```
x x
x x x
  x x
```
10.
```
  x
x x x
x x x
  x
```
11.
```
  x x
x x x x
x x x x
  x x
```

For an s-irreducible set only configurations 1, 2, 3 and 7 are possible as kernel components.

For a proof of the Theorem see |2|, Theorem 5.1 and |4|.

A detailed discussion of the method together with realistic examples is presented in a separate report |6|.

References

1. Arcelli, C.: Pattern thinning by contour tracing. Computer Graphics and Image Processing 17 (1981) 130-144.
2. Eckhardt, U. and Maderlechner, G.: Parallel reduction of digital sets. Siemens Forsch.- u. Entwickl.-Ber. 17 (1988) 184-189.
3. Eckhardt, U.: A note on Rutovitz' method for parallel thinning. Pattern Recognition Letters 8 (1988) 35-38.
4. Eckhardt, U. and Maderlechner, G.: The structure of irreducible digital sets obtained by thinning algorithms. 9th ICPR, Rome, Nov. 14-17, 1988, Proceedings, pp. 727-729. Washington IEEE 1988.
5. Eckhardt, U.: Anwendung der digitalen Topologie in der Binärbildverarbeitung. Mitt. Math. Ges. Hamburg XI (1989) 6, 719-736.
6. Eckhardt, U. and Maderlechner, G.: Thinning of binary images. Hamburger Beiträge zur Angewandten Mathematik, Reihe B, 1989.
7. Eichhorn, B.: Skelettierung mit Randinformation. Diplomarbeit am Institut für Informatik, TU München, 1987.
8. Favre, A. and Keller, H. J.: Parallel syntactic thinning by recoding of binary pictures. Computer Vision, Graphics, and Image Processing 23 (1983) 99-112.
9. Kreifelts, T.: Skelettierung und Linienverfolgung in rasterdigitalisierten Linienstrukturen. Informatik-Fachberichte 8 (1977) 223-231.
10. Minsky, M. and Papert, S.: Perceptrons. An Introduction to Computational Geometry. Expanded Edition. The MIT Press, Cambridge 1988.
11. Ogawa, H. and Taniguchi, K.: Thinning and stroke segmentation for handwritten Chinese character recognition. Pattern Recognition 15 (1982) 299-308.
12. Ronse, C.: Minimal test patterns for connectivity preservation in parallel thinning algorithms for binary digital images. Discrete Applied Mathematics 21 (1988) 67-79.
13. Rosenfeld, A.: Digital topology. Amer. Math. Monthly 86 (1979) 621-630.
14. Stefanelli, R. and Rosenfeld, A.: Some parallel thinning algorithms for digital pictures. J. Assoc. Comput. Machinery 18 (1971) 255-264.
15. Tamura, H.: A comparison of line thinning algorithms from digital geometry viewpoint. 4th ICPR, Kyoto, Nov. 7-10, 1978, Proceedings, pp. 715-719. Washington IEEE 1978.
16. Xia, Y.: A new thinning algorithm for binary images. 8th ICPR, Paris, Oct. 27-31, 1986, Proceedings, pp. 995-997. Washington IEEE 1986.

"DIGITAL GEOMETRY" ON GRAPHS

Azriel Rosenfeld[1]
Angela Y. Wu[2]

ABSTRACT

Many of the standard concepts of digital geometry, particularly those involving connectedness and distance properties of subsets of a digital image, can be generalized to subgraphs of an arbitrary graph G. Algorithms for connected component labeling, distance transform computation, etc. can be defined that require time (e.g.) $O(n+a)$, where n is the number of nodes of G and a is the number of arcs. Parallel algorithms for these computations can also be defined, using various models of parallel computation. We can also define "continuous" integer-valued functions on graphs, and can show that the distance transform is the largest such function having (at least) a given set of zeros.

1. Introduction

Digital geometry deals with geometric properties of subsets S of digital images [6]; among these properties are

a) *Connectedness.* Pixels P,Q of S are said to be *connected* in S if there exists a sequence of pixels $P=P_0, P_1, \ldots, P_m=Q$ of S such that P_i is adjacent to P_{i-1}, $1 \leq i \leq m$. The maximal connected subsets of S are called its connected ***components.*** It is often useful to *label* the connected components of S, by assigning a (nonzero integer) value to each pixel of S such that two pixels have the same value iff they belong to the same component.

b) *Distance.* The *distance* between pixels P,Q is the length of the shortest sequence of pixels $P=P_0, P_1, \ldots, P_m=Q$ such that P_i is adjacent to P_{i-1}, $1 \leq i \leq m$. It is often useful to compute the distance from each pixel of S to the nearest pixel of $\overline{S}$. This assigns a nonzero integer value to each pixel of S; the resulting array of values is called the "distance transform" of S. The *intrinsic distance* between pixels P,Q of S is the length of the shortest sequence of pixels $P=P_0, P_1, \ldots, P_m=Q$ *of* S such that P_i is adjacent to P_{i-1}, $1 \leq i \leq m$.

c) *Area, parameter, diameter.* The *area* if S is simply the number of pixels in S. The *perimeter* of S is the number of pixels of S that are adjacent to pixels of $\overline{S}$. The *diameter* (*intrinsic diameter*) of S is the greatest distance (intrinsic distance) between any two pixels of S.

The pixels of a digital image are usually assumed to correspond to lattice points (i.e., points having integer coordinates) in the plane; but the geometric concepts described above can be defined for other types of discrete arrays of points—for example, hexagonal arrays in the plane, or arrays of lattice points in 3-space. Indeed, these concepts depend only on having a well-defined notion of "adjacent". Note that for a given type of array there may be more than one such notion. For example, in the case of lattice points in the plane we can define P,Q to be adjacent if they are horizontal or vertical neighbors (i.e., if $P=(x,y)$ and $Q=(u,v)$, then $|x-u| + |y-v|=1$), or we can also allow them to be diagonal neighbors ($\max(|x-u|, |y-v|) = 1$). Different notions of adjacency give rise to different concepts of connectedness, distance, etc.; for

[1]Computer Vision Laboratory, Center for Automation Research, University of Maryland, College Park, MD 20742-3411
[2]Department of Computer Science and Information Systems, American University, Washington, DC 20016

example, for lattice points in the plane, if only horizontal and vertical neighbors are regarded as adjacent, "distance" means "city block distance" $\left(d\left((x,y),(u,v)\right) = |x-u| + |y-v|\right)$, and if diagonal neighbors are also regarded as adjacent, it means "chessboard distance" $\left(d\left((x,y),(u,v)\right) = \max(|x-u|,|y-v|)\right)$.

Since the concepts of connectedness, distance, etc. depend only on the notion of adjacency, they can be defined not only for subsets of arrays, but for subsets of (the nodes of) a graph, where two nodes are adjacent iff they are joined by an arc. In Section 2 of this paper we give these definitions. In Section 3 we give algorithms for labeling the connected components of a given subset S of the nodes, and for computing its distance transform, in time $O(n+a)$, where n is the number of nodes of the graph and a is the number of arcs. Section 4 discusses various types of parallel algorithms for these computations. Finally, in Section 5 we consider "continuous" functions on the nodes of a graph (integer-valued function whose values at adjacent nodes differ by at most 1) and show that the distance transform of S is the largest such function that takes nodes of $\bar{S}$ into 0.

2. Definitions

Let G be a graph. We call two nodes P,Q of G *adjacent*, or *neighbors*, if they are joined by an arc. We call P,Q *connected* if there exists a sequence of nodes $P = P_0, P_1, \ldots, P_m = Q$ such that P_i is adjacent to P_{i-1}, $1 \leq i \leq m$. Such a sequence is called a *path* from P to Q. The maximal connected subsets of (the nodes of) G are called its *connected components*. If G consists of a single connected component, it is called connected.

The *distance* from P to Q, denoted $d(P,Q)$, is the length of the shortest path from P to Q. Evidently, $d(P,Q)$ is a nonnegative integer valued metric, and is defined only if P and Q are connected. (We recall that a metric has the properties $d(P,Q)=0$ iff $P=Q$; $d(P,Q)=d(Q,P)$; and $d(P,R) \leq d(P,Q) + d(Q,R)$, for all nodes P, Q, R of G.)

Let S be a subset of the nodes of G; then S can be regarded as a subgraph of G, where two nodes of S are joined by an arc in S iff they are joined by an arc in G. We call nodes P,Q of S connected in S if they are connected in the subgraph S; connected components of S and connectedness of S are defined correspondingly.

The distance transform of S assigns to each node of S its distance to the nearest node of $\bar{S}$. (Note that it is defined only if there are nodes of $\bar{S}$ in each connected component of G.) The intrinsic distance between nodes P,Q of S is their distance in the subgraph S; evidently this is at least as great as (and may be greater than) their distance in G.

The area, perimeter, and (intrinsic) diameter of S are defined just as in Section 1, with "pixels" replaced by "nodes". (Note that the perimeter is defined only if $\bar{S}$ is nonempty.)

3. Sequential algorithms

Let S be identified by labeling the nodes of G that belong to S with 1's and those that belong to $\bar{S}$ with 0's, as is conventionally done to identify a subset S of a digital image. For brevity, we shall sometimes refer to the nodes of S and $\bar{S}$ as 1's and 0's, respectively.

To find the area of S we need only examine every node of G and count the 1's. Evidently this requires $O(n)$ steps, where n is the number of nodes of G. To find the perimeter of S we examine each node of G, and when we find a 1, visit all its neighbors to determine if any of them is a 0. This requires $O(n+a)$ steps (strictly, $n+a'$ where a' is the number of arcs emanating from the nodes of S). These algorithms are analogous to the standard ones for finding the area and perimeter of a subset of a digital image.

To find the (intrinsic) diameter of S, we can use Dijkstra's algorithm [1] to find the distance from each node P of S to every other node of S in the graph G (or in the induced subgraph S, in the case of intrinsic diameter), and the (intrinsic) diameter is the maximum of these values. Dijkstra's original algorithm is $O(n^2)$. Thus finding diameter is $O(mn^2)$, where m is the number of nodes of S; and finding intrinsic diameter is $O(m^3)$. Using a heap and adjacency list, the time complexity of the Dijkstra algorithm is $O(a \log n)$ [2]. Hence finding the diameter of S takes time $O(ma \log n)$ for diameter; $O(ma' \log n + n)$ for intrinsic diameter. Improvements on these times would be desirable.

The standard method of labeling the connected components of a subset of a digital image [7] is as follows: (a) Traverse the image, e.g. row by row. (b) When a 1 is found (the "current 1"), examine its neighbors that have already been visited by the traversal. (c) If none of them is 1, give the current 1 a new label. (d) If some of them are 1's and they all have the same label, give the current 1 that label. (e) If some of them are 1's and they have more than one label, give the current 1 one of these labels, and note that these labels are "equivalent". (f) After all equivalences have been found in this way, choose a representative of each equivalence class and replace all labels in the class by its representative. A similar algorithm can be used to label the connected components of a subset of a graph G, by traversing G (e.g. construct a breadth-first directed spanning tree of G, and traverse the tree in some standard order). Evidently, this algorithm requires $O(a' + n)$ steps plus the time required to choose the representatives of the equivalence classes. [Note that if G itself is a tree, and the traversal goes from neighbor to neighbor, no equivalences can ever be discovered; if the current node has two neighbors that have both already been visited, they are connected by a path that does not pass through the current node, so that G has a cycle, contradiction.]

The equivalences can be managed using the standard Union and Find algorithms [8], where each equivalence class is represented by a tree. In Step (e), when two labels are equivalent, Find operations are performed to find the equivalence class representatives of these labels (the roots of the trees) and then a Union operation is performed to merge the two trees. These operations take time proportional to the height of the tree, which is $\leq \log l$ where l is the number of labels in the tree. This time bound can be improved if path compression is performed during the Find operation [9].

Another way to label the connected components of a subset S of the nodes of a graph G is to start with a node in S (with label 1); label all the 1's connected to it (treating the arcs (of G) between 0 and 1 nodes as nonexistent); then take another node of S which is not labeled and label its component; and so on until all nodes of S are labeled. To do this, we need to keep a list of the nodes of S which are labeled but have neighbors that are not yet labeled; but no equivalences of labels are needed. This algorithm takes $O(a' + n)$ time. Note that in this algorithm, the order of traversal of the nodes is determined by the 1's (the S nodes), while in the first algorithm, the order of traversal is the same for any S.

The standard method of computing the distance transform of a subset of a digital image [7] is as follows: (a) Traverse the image, say row by row, top to bottom. (b) When a 1 is found, give it a value 1 greater than the minimum of the values of its neighbors that have already been visited by the traversal. (c) Traverse the image in reverse order (bottom to top). (d) When a nonzero value is found, give it a new value 1 greater than the minimum of the values of its neighbors that have already be visited by the traversal (but not less than its old value). Evidently this algorithm requires $O(n)$ steps, where n is the image area. It is not clear whether an analogous algorithm can be devised for computing the distance transform of a subset of a graph. However, we can compute it using the following algorithm:

(1) Put all the 1's that have 0's as neighbors into a queue and give them distance values of 1.

(2) While the queue is nonempty, take the first node from the front of the queue, say node P with value d; find all the neighbors of P that have not yet been given values, give them value $d+1$ and put them at the rear of the queue.

Evidently this algorithm requires $O(a'+n)$ steps. If in step (2) when a node is given value $d+1$, it also remembers node P, then a path to the nearest zero can be constructed by following this information in the nodes.

If the arcs of G do not have constant weight, then the above algorithm does not work. However, by adding to G a special node joined by arcs of zero weight to all the 0's having 1's as neighbors, we can use Dijkstra's single source all shortest path algorithm to find the shortest distance from this extra node to all the 1 nodes. Clearly, this gives the distance to the closest 0 from each node of S in $O(n+m^2)$ or $O(n+a'\log m)$ time depending on the data structures used to implement the algorithm. (Here the n term is the time required to find the nodes that need to be joined to the new node.)

4. Parallel algorithms

We can also define parallel algorithms for performing the geometric computations discussed above, using various models of parallel computation. In this section we discuss "cellular graph" algorithms [10]; we will not consider algorithms for other architectures, e.g. mesh, hypercube, shared memory, etc. (see [4]).

Assume that we have a set of synchronized processors corresponding to the nodes of G, and that two processors can communicate directly iff the corresponding nodes are joined by an arc. Thus this multiprocessor network has the structure of G, and we can give it the definition of S by simply inputting 0's and 1's to the appropriate processors. A variety of algorithms for such processor networks are given in [10], where it is assumed that G has bounded degree and that the processors are unite-state automata.

If we choose one of the processors, say P_0, as a root, we can mark the arcs that define a breadth-first directed spanning tree rooted at P. The root node can then compute the area of S in O(tree height) steps as follows: Each leaf node counts 0 or 1 depending on whether its value is 0 or 1; each non-leaf node, when all its children have computed their counts, adds these counts. The perimeter of S can be computed similarly, with each leaf node counting 1 provided it has value 1 and has a neighbor with value 0. It would be of interest to define efficient algorithms for computing the (intrinsic) diameter; see [10], which gives a slow algorithm for computing the "radius" (the minimum, over all nodes P, of the maximum distance from P to any other node).

To label the connected components of S, each node having value 1 gives itself a distinct numerical value (its node number), and each node then repeatedly changes its value to (say) the minimum of its neighbors' values. Evidently, this gives all the nodes in a given component C of S the same value after a number of steps equal to the intrinsic diameter of C.

To compute the distance transform of S, each node having a non-zero value repeatedly changes its value to 1 greater than the minimum of its neighbors' values. Evidently, this yields the distance transform of a given component C of S after a number of steps equal to the radius of

C, i.e. the greatest distance of any node of C from $\bar{S}$.

5. "Continuous" functions on graphs

In [5] we defined a function from an array into an array as being "continuous" if it takes adjacent pixels into the same or adjacent pixels. More generally, we can define a function from the nodes of a graph G into the nodes of a graph H as being "continuous" if it takes adjacent nodes into the same or adjacent nodes. In terms of graph distance, this is equivalent to requiring that $d(P,Q) \leq 1$ implies $d(f(P), f(Q)) \leq 1$. Similarly, a function from the nodes of a graph into a metric space will be called "continuous" if $d(P,Q) \leq 1$ implies $d'(f(P), f(Q)) \leq 1$, where d' is the metric on the space. In either case, we call f *nonincreasing* if $d^{(\prime)}(f(P), f(Q)) \leq d(P,Q)$.

Proposition 5.1. *f is nonincreasing iff it is continuous.*

Proof. "Only if" is clear. We prove "if" by induction on $d(P,Q)$; "if" is the base case $d(P,Q)=1$. Suppose the assertion true for $d(P,Q) \leq k$, and let $d(P,Q)=k+1$; then there exists R such that $d(P,R)=k$ and $d(R,Q)=1$. By induction hypothesis, $d^{(\prime)}(f(P), f(R)) \leq k$, and by the base case, $d^{(\prime)}(f(R), f(Q)) \leq 1$; hence by the triangle inequality, $d^{(\prime)}(f(P), f(Q)) \leq k+1$.//

Note that in particular, when f is a function from G into H, and P,Q are connected in G (so that they have a finite distance), $f(P)$, $f(Q)$ must be connected in H; thus if $f\colon G \rightarrow H$ is continuous, it takes connected subgraphs into connected subgraphs. Conversely, if f takes two-node connected subgraphs (i.e., adjacent pairs of nodes) into connected subgraphs, it must take adjacent nodes into the same or adjacent nodes. We thus have

Proposition 5.2. *$f\colon G \rightarrow H$ is continuous iff it takes connected subgraphs in connected subgraphs.*//

Let G be a graph with nodes labeled 0 and 1, say with set of 1's S. The distance transform of G is a nonnegative integer valued function $d_{\bar{S}}$ defined on the nodes of G. If P,Q are adjacent nodes, their distances to $\bar{S}$ differ by at most 1; hence $d_{\bar{S}}$ is continuous.

Proposition 5.3. *$d_{\bar{S}}$ is the largest continuous integer-valued function defined on the nodes of G and having set of zeros $\supseteq \bar{S}$.*

Proof. Let d be any such function; we must show that $d_{\bar{S}}(P) \geq d(P)$ for all P. This is clear if $d_{\bar{S}}(P)=0$; we proceed by induction on $d_{\bar{S}}(P)$. Suppose it is true for all $d_{\bar{S}}(P) \leq k$, and let $d_{\bar{S}}(P)=k+1$. Then P has a neighbor P' such that $d_{\bar{S}}(P')=k$. By induction hypothesis we have $d(P') \leq k$, and by continuity of d we have $d(P) \leq d(P')+1$, so that $d(P) \leq k+1$.//

6. Concluding remarks

Many of the standard concepts of digital geometry, including connectedness and distance, depend only on the notion of adjacency between two pixels, and so are defined for subsets of a graph. The standard array algorithms, e.g. for connected component labeling and distance transform computation, do not generalize to graphs, but these computations can still be performed in $O(n)$ time.

Other concepts used in digital geometry do not seem to have natural generalizations to graphs—convexity, for example (but see [3] for directed graphs). These concepts seem to have natural definitions only for arrays that are embedded in Euclidean space.

References

1. E.W. Dijkstra, A note on two problems in connexion with graphs, *Numerische Mathematik* **1**, 1959, 269–271.
2. D.B. Johnson, Efficient algorithms for shortest paths in sparse networks, *J. ACM* **24**, 1977, 1–13.
3. J.L. Pfaltz, Convexity in directed graphs, *J. Combinatorial Theory* **10**, 1971, 143–162.
4. M.J. Quinn and N. Deo, Parallel graph algorithms, *ACM Computing Surveys* **16**, 1984, 319–348.
5. A. Rosenfeld, "Continuous" functions on digital pictures, *Pattern Recognition Letters* **4**, 1986, 177–184.
6. A. Rosenfeld and A.C. Kak, *Digital Picture Processing*, Academic Press, New York, 1982, Chapter 11.
7. A. Rosenfeld and J.L. Pfaltz, Sequential operations in digital picture processing, *J. ACM* **13**, 1966, 471–494.
8. R.E. Tarjan, Efficiency of a good but not linear set union algorithm, *J. ACM* **22**, 1975, 215–225.
9. R.E. Tarjan and J. Van Leeuwen, Worst-case analysis of set union algorithms, *J. ACM* **31**, 1984, 245–281.
10. A.Y. Wu and A. Rosenfeld, Cellular graph automata, *Information Control* **42**, 1979, 305–353.

ALGEBRAIC THEORY OF DISCRETE SIGNALS

K. Voss, C. Hermann [1])

It is possible and necessary too to develop a logically closed theory of digital image processing resulting from extensive applications of discrete algorithms in both general-purpose computers and specially constructed digital processing devices. The classical theory of analog signal processing systems is analysis. Many algorithms and techniques in digital signal processing, among which is image processing, only approximate analog techniques. However, there are some discrete algorithms having no analogue in analog signal processing. Indeed, the fast Fourier transform algorithm (FFT) can not be applied for computing Fourier integrals. The calculation of a finite digital convolution makes use of algebraic methods based on Chinese remainder theorem, number theoretic transforms (NTT), and discrete convolution theorem (DCT), which have nothing to do with analysis. That means, digital signal processing systems can not be subordinated to analog signal processing theories, nevertheless, they are a reality with own properties and own laws. Consequently, the structural characteristics of discrete phenomenons must be the starting point to develop a theory of digital signal processing. First, we have to replace analysis by algebra. Secondly, the finite character of all numerical algorithms suggests using finite structures and sets for the base principles to produce an algebraic theory of discrete signals.

1. Translative neighborhood structures

The basis of a digital image is a lattice structure. The lattice structures serve as signal carrier of digital images regarded as digital signals. Over the past few years, generalized models of suitable lattice structures are developed (/2/, /3/, /9/, /11/). Such models we call neighborhood structures.

Definition 1: Let P be a finite set of points and let $N \subset P \times P$ be a symmetrical and irreflexive relation. Then the tuple $[P,N]$ is called neighborhood structure , and two elements p and q of P are called neighbors if $(p,q) \in N$.

It is easily provable that neighborhood structures are isomorphic to undirected graphs without loops and multiple edges (graphs of neighborhood). In a natural way, graphs of neighborhood allow to define a metric: The distance $d(p,q)$ between points $p,q \in P$ is equal to the minimal length of all paths connecting p and q counting every edges of graph with length one. We regard neighborhood structures as the discrete analogues of topologies of uncountable spaces. A one-to-one mapping α from a graph onto itself such that neighborhood relation N is not changed, $\alpha(N)=N$, is called discrete topological mapping.

Definition 2: A special discrete topological mapping is called translation if the identity

$$d(p,\tau(p)) = d(q,\tau(q))$$

for all $p,q \in P$ is valid.
Generally, the set T of all translations of a given neighborhood is not a group (/2/, /3/).

Definition 3: $[P,N]$ is called a translative neighborhood structure if the following conditions are satisfied:

1) The set T of all translations represents a commutative group G.

---

[1]) Friedrich Schiller University, Dept. of Technology,
E. Thaelmann-Ring 32, Jena 6900, DDR

2) For any two points $p,q \in P$ exists one and only one translation $\tau \in G$ mapping p in q, $\tau(p)=q$.

Furthermore, $[P,N]$ is called a weak translative neighborhood structure if G is only a subset of T, $G \subset T$. From now we shall regarded $[P,N]$ as weak translative. If we fix an arbitrary point $o \in P$, the mapping τ --> p, $p=\tau(o)$, is a one-to-one mapping from G onto P. Denoting $\tau(o)=p$ by τ_p, we induce in P an additive group structure p+q defined as followes

$$\tau_{p+q} = \tau_p \tau_q, \quad p,q \in P. \tag{1}$$

Particularly, τ_p^{-1} corresponds to -p and $\tau_p \tau_q^{-1}$ corresponds to p-q, respectively. In the following, we replace the translation group G by the group structure (1) in P. Then the order O(P) of the group P is finite, because the cardinality of P is finite. If

$$P = P_1 \oplus \dots \oplus P_r$$

is the factorization in the direct product of cyclical subgroups $P_1,\dots,P_r$ with prime power order, then the order O(P) is given by $O(P)=O(P_1)\ \dots\ O(P_r)$ and n denotes the least common multiple of $O(P_1),\dots,O(P_r)$.
It is clear that the cyclical subgroups $P_1,\dots,P_r$ are connected with the periodicities of the weak translative neighborhood structure $[P,N]$.

2. Signals and signal spaces

In digital signal processing is neither differentiated nor integrated. However, digital signals can be added, subtracted and multiplied. General divisions do not occur. The finite resolution of digital signal processing systems, together with the exclusion of general division, suggests using (finite) rings, fields and algebras. This automatically raises the question: are these algebraic structures suitable tools to found a digital signal theory?
The theories of discrete Fourier transform (DFT) and the FFT algorithm were investigated in finite fields and rings by NICHOLSON /4/ and POLLARD /6/. The FFT algorithm applications to Fourier, Walsh and Hadamard transforms were shown to be a special case of Fourier transform in algebra over fields or rings. Over the past few years, attention has been focused on NTT's over modulo rings Z(m), where m is a positive integer (/1/, /7/). In the recent development of algorithms for digital signal processing polynomial rings have played a crucial role (/5/, /8/).
The common goal of these research is characterized by the development of fast algorithms and transforms and the construction of corresponding computer architectures. In the present papers we want to generalize the algebraic approach for computing circular convolutions based on above mentioned investigations independent of special digital machines. A first project was introduced in /10/.
Let $[P,N]$ be a weak translative neighborhood structure.

Definition 4: The set S of all unique mappings f: P ---> R from P of $[P,N]$ into a commutative ring R with identity is called discrete signal space and its elements are called discrete signals.

The assumption, R is a commutative ring with identity, is not necessary but sufficient and practical. In each commutative ring with identity 1, there exists a subring of integers $Z(m)=\{0,1,2,\dots,m-1\}$ where m is the characteristic of the ring. In general, the output values from a digital signal processing system are sequences of integers (discrete signals). If m is greater than the maximum output value, then there are enough elements in this subring to represent all necessary values. Hence, calculations on ordinary integers can be translated into ring operations.
More algebraically, R is called a domain of integrity if for each nonzero element a of R there exists no nonzero element $b \in R$, such that $ab=0$ (0 is the null element of R). For instance, the integers, reals and the complex number field are domains of integrity.

Given the preceding definitions and the trivial definition

$$(f+g)(p) = f(p) + g(p), \quad p\in P,$$
$$(af)(p) = af(p), \quad p\in P,\ a\in R, \tag{2}$$

where f and g are elements of a discrete signal space S we structure S to a finite dimensional vector space. The discrete "delta" functions

$$e_q(p) = 1 \quad \text{for } p=q,\ p\in P$$
$$= 0 \quad \text{for } p\neq q,\ p\in P \tag{3}$$

subscripted by $q\in P$ represent a base in S. Indeed, it can be derived from (3) that the equation

$$f = \sum_{q\in P} a_q e_q \tag{4}$$

has a unique solution, that means the sequence a_q, $q\in P$, of elements of R is uniquely determined (in our case $a_q=f(q)$, $q\in P$). In particular, we call the set e_q, $q\in P$ of discrete "delta" functions canonical base. Denote N the number of elements of P, there are N basic elements e_q, $q\in P$, and the dimension of S is equal N, dim(S)=N.
Furthermore, we introduce a new operation, $f*g$, the circular convolution, between elements f and g of S by

$$(f*g)(p) = \sum_{q\in P} f(p-q)g(q), \quad p\in P. \tag{5}$$

The circular convolution is an operation in S, that means all constructions $f*g$ are elements of S and have the following properties

$$(f*g)*h = f*(g*h),$$
$$f*g = g*f,$$
$$f*(g+h) = f*g + f*h,$$
$$a(f*g) = (af)*g, \tag{6}$$

where $f,g,h\in S$ and $a\in R$. Together with the linearity (2), from the circular convolution properties (6) follow that S is a commutative algebra over the ring R. Suppose

$$f = \sum_{q\in P} a_q f_q, \quad g = \sum_{q\in P} b_q f_q, \quad a_q, b_q\in R \tag{7}$$

are a base representation of two elements $f,g\in S$ with reference to a fixed base f_q, $q\in P$. So, because of (6) and (7), the circular convolution of f and g is given by

$$f*g = (\sum_{p\in P} a_p f_p) * (\sum_{q\in P} b_q f_q)$$
$$= \sum_{p,q\in P} a_p b_q (f_p * f_q). \tag{8}$$

Otherwise, the convolutions $f_p * f_q$ are elements of S and, consequently, must be of the form

$$f_p * f_q = \sum_{r \in P} c^r_{pq} f_r, \qquad c^r_{pq} \in R \tag{9}$$

and we obtain

$$f * g = \sum_{p,q,r \in P} a_p b_q c^r_{pq} f_r. \tag{10}$$

The c^r_{pq} are called structure constants of the algebra S. For a fixedly choosen base f_p, $p \in P$, of S, the structure constants are uniquely determined. For the "delta" base (3) the c^r_{pq} can be derived from

$$c^r_{pq} = e_p * e_q(r) = 1 \quad \text{for } p+q=r$$
$$= 0 \quad \text{for } p+q \neq r. \tag{11}$$

3. Convolution orthogonality

The theory of commutative algebras states that the commutative ring R and structure constants c^r_{pq} uniquely determine a commutaive algera, up to isomorphism. Particularly, the signal space S is a commutative algebra. The group P and the neighborhood structure $[P, \mathcal{N}]$, respectively, establish the structure constants (8), i.e., P and R uniquely determine S.
In a given space S, a change of base elements $f_p \text{-->} f'_p$ induce a change of induce a change of structure constants $c^r_{pq} \text{-->} c'^r_{pq}$. The structure constants are known for the canonical base (11). Otherwise, because (10) the number of general multiplications needed to calculate the convolution $f*g$ of two signals f and g is equal to the number of the nonzero components of c^r_{pq}. In the case of the canonical base we have N^2 nonzero components and multiplications, respectively. However, more simple are following structure constants

$$c^r_{pq} = 1 \quad \text{for } p=q=r$$
$$= 0 \quad \text{otherwise}, \tag{12}$$

requireing only N general multiplications. The corresponding base f_p, $p \in P$, on S, provided it exists, we call orthogonal convolution base (OCB), because then the convolutions of basic elements have the property of "orthogonality"

$$f_p * f_q = d_{pq} f_p, \quad p,q \in P \tag{13}$$

where d_{pq} is the Kronecker delta symbol. In this case, the general convolution (10) is given in a very simple form

$$f * g = \sum_{p \in P} a_p b_p f_p \tag{14}$$

where $f = \sum_{p \in P} a_p f_p$ and $g = \sum_{p \in P} b_p f_p$, respectively, are signals of S represented as to a OCB. Equation (14) is equivalent to the discrete convolution theorem (DCT), that is, the DCT maps circular convolution into pointwise

products. Vice versa, the existence of DCT in S demand the existence of a OCB. An efficient solution to this problem exists, based on the properties of R. We give the following theorem without extensive proof.
Let S (regarded as signal space as for weak translative neighborhood structure $[P,N]$) be a commutative algebra over a domain of integrity R with identity. Furthermore, let the commutative group P be isomorphic to the direct product $P=P_1 \oplus \ldots \oplus P_r$ of r cyclic subgroups $P_1,\ldots,P_r$ with prime power orders $O(P_1),\ldots,O(P_r)$. Finally, n denotes the least common multiple, $n=\mathrm{lcm}(O(P_1),\ldots,O(P_k))$.

Theorem: In S exists a OCB if and only if n is invertible in R and R owns a primitive nth root a of unity such that each expression (a^k-1), $k=1,\ldots,n-1$ has an inverse in R:

$$a^n = 1,$$

$$(a^k-1)^{-1} \text{ exists}, \quad k=1,\ldots,n-1, \tag{15}$$

$$n^{-1} \text{ exists}.$$

The conditions are sufficient for the existence of OCB and general Fourier transform, respectively, and the satisfaction of DCT. Over the past years, this was proven in many papers and the DCT has been employed in a growing number of applications in digital image processing. However, the conditions are necessary too.
In the special case of one dimensional cyclic lattic $[P,N]$ and cyclic group P, respectively, the conditions of the theorem regarding existence of OCB in S are very simple. Let P be a cyclic group of order N, hence, P is isomorphical to Z_N denoting the additive cyclic group modulo N. Finally, define R a domain of integrity with identity.

Theorem: For the existence of a OCB in S the following conditions are sufficient and necessary

1) R possesses a Nth root $a \in R$ of unity, $a^N=1$,

2) $(a^k-1)^{-1}$ exists, $k=1,\ldots,N-1$, (16)

3) N is an unit, i.e., N^{-1} exists.

In the case of existence, the OCB are represented up to isomorphism by the elements

$$f_k(n) = N^{-1}a^{kn}, \qquad n \in Z_N,\ k=0,\ldots,N-1.$$

4. Conclusion

In this paper, a general theory is developed in which many of the currently used convolutions methods, including polynomial transforms and number theoretic transforms, can be discussed. Over the past years, many authors have recognized the connection between DCT, FFT and OCB but these connections is one-to-one as shown in this correspondence. The existence of DCT is only a other formulation than the existence of OCB. OCB's are direct constructions in a discrete signal space S being a commutative algebra over a commutative ring R with identity. Obviously, commutative algebras over R are discrete analogues of continuous functions spaces, used in classical signal theory.

5. References

/1/ Agarwal,R.C., Burrus,C.S.: Number theoretic transforms to implement fast digital convolution. Proc. IEEE 63 (1975) 550-560
/2/ Hübler,A.: Ein abstrakter Zugang zur digitalen Geometrie. In: E.Paulus (Hrsg.): Mustererkennung 1987, 9. DAGM-Sympos. Braunschweig (1987), Springer 1987, 151-155
/3/ Hübler,A., Klette,R.: Theoretische Grundlagen der digitalen Bildver-

arbeitung. VII. Geometrie auf Nachbarschaftsstrukturen. Bild und Ton 40 (1987) 229-235
/4/ Nicholson,P.J.: Algebraic theory of finite Fourier transforms. I.Comput.Syst.Sci. 5 (1971) 524-527
/5/ Nussbaumer,H.J.: Fast Fourier Transforms and Convolution Algorithms. Springer 1981
/6/ Pollard,I.H.: The fast Fourier transforms in a finite field. Math.Comput. 25 (1971) 365-374
/7/ Rader,C.H.: Discrete convolutions via Mersenne transforms. IEEE Trans. C-21 (1972 1269-1273
/8/ Reed,I.S., Truong,T.K.: Complex integer convolutions over a direct sum of Galois fields. IEEE Trans. IT-21 (1975) 657-661
/9/ Voss,K.: Nachbarschaftsstrukturen als Bildträger. In: E.Paulus (Hrsg.): Mustererkennung 1987, 9.DAGM-Sympos. Braunschweig 1987, Springer 1987
/10/ Voss,K.: Theoretische Grundlagen der digitalen Bildverarbeitung. VIII. Eindimensionale diskrete periodische Funktionen. Bild und Ton 41 (1988) 300-307
/11/ Voss,K.: Theoretische Grundlagen der digitalen Bildverarbeitung. Akademie-Verlag, Berlin 1988

A METHOD FOR THE STRUCTURAL RECOGNITION OF NOISY PATTERNS

Fuchs, Siegfried 1)
Kovalevsky, Vladimir A. 2)

It is well known, that certain simple objects may be reliably recognized by decision rules based on probability distributions of some numeric features. Complex patterns, such as parts of graphical drawings may be recognized by analysing the relations between the parts of the objects. A typical technique of this kind is that of the subgraph isomorphism as applied to the region adjacency graph of the given image. This technique was generalized for the case of the isomorphism of cellular complexes and various binary relations /1/. It was successfully used for the recognition of object classes being described by deterministic prototype patterns. An important condition for using this technique is a reliable image segmentation which must provide the detection of all object parts corresponding to the primitives of the prototype.

This technique is however not suitable in the cases when the objects to be recognized have an inherently stochastic structure or the results of the segmentation are subject to stochastic errors. The aim of the present work is to combine the advantages of both the statistical and structural techniques.

In /1/ a solution is suggested fo the task: Find patterns, described by a given set of prototypes, in a given image scene.

Both the prototypes and the image scene should be given as a segmented image.The suggested method is running throw following steps: (1) Description of the prototypes. (2) Description of the image scene. (3) Search for admissible mappings of the prototype description onto the scene description. (4) Decision for only one prototype or the rejection.

The steps (1) and (2) are realized by a structure of zero-, one- and two-dimensional elements, called cells, and bounding relations, based on the theory of cell complexes /2/. There a cell complex is defined by

$C=(E,B,dim)$ with E a set of elements (cells), B a face relation and dim a dimension function.

For the topologic description of a two-dimensional segmented image we need 2-dimensional cells, the regions, 1-dimensional cells, the boundary lines, and branching points as Ø-dimensional cells. The resulting topologic descriptions of such an segmented image and each prototype are block complexes.They are implemented by a data structure called cell lists.These lists contain additional to the topology metric information and quality attributes. An example is given in Fig.1 for a simple image. The step (3), the search for admissable mappings is controlled by 1- and 2-dimensional predicates PR1 and PR2 assigned to the cells of the prototypes and pairs of them and being avaluated with the mapped cells of the image scene as arguments. Finally the admissibility of a mapping is decided by a global predicate GP, calculated from the set of all predicates, assigned to the mapped prototype. For an admissible mapping MC the global predicate must be TRUE. Finally a logical decision rule based on answers of all global predicates generates an unambiguos decision. The method is shown in Fig.2a as a block diagram.

The checking of the global predicate may be interpreted as the recognition of that object class, which is represented by the prototype. But the checking of the 1-place predicates may be seen as the recognition of classes of details.

The described method works well, as demonstrated in /1/ under the following conditions: - The details may be recognized with a high reliability and undependently from each other, and - the objects are composed from details in that variants , for which prototypes may be stored and - the prototypes are stongly assigned to pattern classes.

We may overcome this drawback, if we replace the decisions TRUE or FALSE e.g. by probability maesures or fuzzy functions or feature values only. These proposed alterations generalize the above described method replacing the predicates by mappings of the cells or cell pairs onto the set of real numbers, e.g. the so called features or probability measures. The problem in the generalized form is shown in Fig.2b .

The solution of this problem is the best mapping from the set of prototypes onto the image scene, the best e.g. in the bayesian sense. This solution can be found by an exhaustive search. But this is not acceptable in real problems and the advantages of the formerly suggested method may be lost.This drawback can be avoided under the

condition, that in each class is at least one detail, which can be recognized with a high reliability. These key elements can be used as start points for the mapping process in the following manner:
A key element is represented in the prototype list by a cell, e.g. a region, and an assigned one-place predicate (key predicate). In the list of the image block complex that regions are searched, which satisfy these predicates. As it is shown in the recognition scheme (Fig.3), all mappings of the prototypes, which may be started at any key will be put in a list L ; for each mapping a line with the prototype name, a running number of the mappings, the set of cells onto which the prototype is mapped, the discriminant function of the maximal likely decision and the corresponding decision number. Then all pairs of lines in L will be searched with mapped subcomplexes of the image intersecting eachother. That line of each pair with the less discriminant function (less likely decision) is deleted from L. At the end of this process we will get in L the list of uniquely and optimal mappings and recognized objects. A further reduction of the search may be achieved using key relations between pairs of elements.

The suggested generalized method shall be demonstrated by a model example, constructed in a simplified manner for understanding the handling of the method only.
Given the task: An automaton has to decide between two species of fishes:

M mackerel H herring

The recognition shall be based on pictures from the side Fig.4. About this task may be known:

- A feature vector $\underline{x}=(x1\ x2)$ describing the shape of the body regions $x1=(b+c+d)/a$ $x2=(b+d)/c$ (for a,b,c,d see Fig.4)
- We assume the feature vector of each class to be normal distributed with given parameters.
- The apriori probabilities P(KM)= Ø,4 and P(KH)= Ø.6 .
- The remarkable structural detail: A STRIPE pattern.- And the machine may detect this pattern with the probabilities P(STRIPE| KM)= Ø.75 and P(STRIPE|KH)= Ø.1 .

We shall show a solution with the bayesian decision rule

$$\hat{d} = \arg\max_{k \in \{KM,KH\}} DFk(STRIPE,\underline{x})$$

with

$$DFk(STRIPE,\underline{x}) = [\ln P(k) + \ln P(STRIPE|k) - ((\underline{x}-\mu k)^2/2\sigma^2)]$$

This rule generates a pair of linear functions, separating the feature space each into two decision regions (dividing functions A and B Fig.5) and the structural predicate STRIPE determines which of them is in action. The rule may be realized by the suggested method and we will describe the solution in terms used in Fig.2b . In the following we call this method 1 . According to Fig.2b the following components are needed:

(1) Prototypes: We construct prototype images, which contain no more details than needed for the destinction between the two classes (Fig.6a) and we describe these prototype images by cell complexes in form of cell lists.

(2) Predicates and mappings assigned to the cells of the prototypes (Fig.6b). First of all we determine a predicate, which corresponds to a remarcable detail (key) of the objects. In our case PR 1Ø describing the typical shape of the body silhouettes of fishes. This predicate is calculated for all regions of the image complex and it is expected to be TRUE for the body region only to start the mapping process at region R2 (Fig.6b). Then the terms of the discriminant function handled in the decision rule given above have to be calculated on the features of those regions of the image complex onto which the regions of the prototype complex are mapped. These calculations are mappings from the subset of region cells into the domain of real numbers. For one of the two classes the mapping M11 respectivly M12 is calculated with the corresponding mean vector $\underline{\mu}k$ and the shape feature vector $\underline{x}$ measured on the mapped body region. Further the probability terms M21 and M22 are determined based on the STRIPE predicate, which will be TRUE if at least two stripe shaped regions (Ri,Rj) are in parallel and with a distance smaller than a threshold.

(3) Assignments (Fig.6b): For each prototype exists a table assigning predicates or other mappings to those cells of the prototype for which they are calculated only.

(4) Discriminant funtions: Bayesian discriminant functions are calculated on the outputs of the assigned predicates and mappings for all classes in question and for all prototypes which are mapped on parts of the image complex intersecting at least one cell of an other prototype. In our example: When the predicate P1Ø is TRUE for any region, than this is a candidate for the body region R2 both of the

prototypes P1 and P2. For that reason the discriminant functions of both prototypes and both classes are to be calculated.

(5) Decision rule: The decision rule gives finaly the result comparing the discriminant functions between classes and prototypes.

Now we will compare the suggested method with alternatives, which are simpler in the sense that they use some parts of information only.

Method 2 :

The method 2 shall decide using the bayesian rule

$$\hat{d} = \arg \max_{k \in \{KM, KH\}} P(k)\, p(\underline{x}|k)$$

based on the shape feature vector only (not on the STRIPE pattern), but with the same probability distribution. In this case the linaer deviding function C (Fig.5) is generated. Method 2 is the classical way to recognize objects using numerical features and the bayesian rule.

Method 3 :

An other alternative shall use an hierarchical classifier with two stages. The first stage decides by the structural feature STRIPE only. When the STRIPE is TRUE than the final decision will be "mackerel", else the decision is made on the shape teature vector $\underline{x}$, but using the conditional probability P(k|not STRIPE) instead of P(k) as apriori probability. This second stage classifier realizes the dividing function A (Fig.5).

Method 4 :

Finally we will compare our suggested method with the use of the structural feature STRIPE only.

From the theoretical comparison of the four described methods follows, that the full information given in the task statement is used in the method 1 only. The decision in method 2 is based on the shape feature vector only and the structural scene description supports the finding of the body region. In the method 3 both the structural feature STRIPE and the shape feature vector is used, but the uncertainty consisting in the class prediction by the STRIPE feature is ignored. And finally the method 4 decides using the structural STRIPE feature only, ignores the information of the shape feature vector and the uncertainty at all. The above given assumptions of the probability distribution, feature dependences and so on gives the ability to calculate the probabilities of the false decision. The Tab.1 shows the results. They are in accordance with the theoretical expectations: The best results are given by the suggested method 1 .

We have shown a method for the recognition of objects in complexly structured image scenes.This method is based on a formerly suggested one /1/, which was known powerfully handling deterministic image structures; and it is now generalized to the use of probabilistic or other fuzzy described problems.

/1/ Kovalevsky V.A.:Object recognition by structural image analysis, Computing,Springer Vienna 1989 (to be printed)

/2/ Kovalevsky V.A.: Finit topology as applied to image analysis, CVGIP,1989 (to be printed)

Tab.1 : Results in terms of failure probabilities

Method	characterization	failure probability
1	using structural and numericalteatures and probability maesures	8.9 %
2	using numerical features and probability measures	13.9 %
3	hierarchical classifier first stage: (deterministic) structural feature second stage: (probabilistic) numerical features	11.6 %
4	deterministic classifier using structural feature only	16.0 %

----

1) Prof.Dr. Siegfried Fuchs, informatik-Zentrum TU Dresden, Mommsenstr.13 Dresden DDR-8027

2) Prof.Dr. Vladimir A. Kovalevsky, Zentralinstitut fuer Kybernetik und Informationsprozesse der AdW der DDR, Kurstr.33, Berlin DDR-1086

The figures to this paper are at pages 249 - 253

LOCAL CRITERIA : A UNIFIED APPROACH TO LOCALLY ADAPTIVE LINEAR AND RANK FILTERS

L.P.YAROSLAVSKY
Institute of Information Transmission Problems
Academy of Sciences of the USSR.
101447 Moscow,GSP-4,Yermolovoy str.19

During the last several years a lot of new linear and nonlinear rank algorithms for signal and picture processing have been proposed [1-7].The linear ones have been derived from one or another statistical model of the signal to be processed.The rank ones have been substantiated from different positions and without any rather general notion of optimality.The aim of this paper is to introduce a new family of criteria of processing quality as a basis for unified approach to synthesis of optimal linear and rank filters for signal and picture processing.

Any reasonable criterion of signal processing quality,i.a. of estimating signal and making decision through its observed realization assume measurement of estimation quality on the average,i.a. averaging of somehow defined function of losses due to the discrepancy between the estimated and desired ("true") signal values,over some set.Presently,most popular is statistical treatment of the set over which the loss function is averaged which means that the processed signal is regarded as a sample function of some statistical ensemble.In this connection,studies oriented to statistical picture models are abundant in the picture processing field,and the expectation of "good" processing algorithms are connected with development of a "correct" statistical model.If method of processing fails,people think that the model is inappropriate.It is the present author's opinion that the true reason for algorithms to be good or bed lies in processing criteria rather than in signals statistical models. Criterion and statistical models are in essence inseparable,the former been primary because it defines the terms in which the model is to be constructed.The model is part and parcel of the criterion,in a sense.In the following we shall discuss this problem primerily for the picture processing tasks.

When formulating picture processing criteria one should proceed from the fact that the resulting picture is not the end product but rather a subject for subsequent (most commonly, visual) analysis.From this standpoint ,two essentially different kinds of criteria such as detail and texture ones may be distinguished defining two kinds of models [13].The former criteria regard pictures as a combination of interpretation objects (details) carrying information in a specific picture analysis application and the background,i.e.the component that in this application carries no information.The latter criteria imply that the picture cannot be splitted into details and background and are to be used if the aim of picture analysis is to determine some parameters of total picture or to classify it as a whole.

When using the texture criteria,one can reasonably regard the picture as realization of some random process,and averaging of loss function as averaging over an ensemble defined by this process.

For the detail criteria,statistical description of picture and,correspondingly,the notion of loss function averaging involved into them are more complicated.

The local processing criteria considered in this paper fall into the class of detail criteria.They feature averaging of loss function over a subset of picture samples in addition to statistical averaging.To elucidate this statement,introduce the following notations.

Let Vin={Vin(m,n)} be vector of input picture samples,m,n - integer coordinates of of samples,Vout={Vout(m,n)} - vector of the resulting output pixels,Vtr={Vtr(m,n)} - vector of "true" (desired) pixels,LOSS(Vtr(m,n);Vout(m,n)) - loss function.The quality measure defining the class of local criteria is then as follows:

$$LOSSav(k,l)=$$

$$AVimsys\{AVob\{AVbgrd\{\sum_{m,n} LOC(m,n;Vtr(k,l))LOSS(Vtr(m,n);Vout(m,n)\}\}\}$$

where (k,l) are the coordinates of pixel for which the averagé loss function is measured,LOC(m,n;Vout(k,l)) - is a so called locality function,i.e weight function that is non-zero for some subset of pixels surrounding the (k,l) one in some sense, AVimsys AVob and AVbgrd are for statistical averaging,respectively,over internal noise of IMaging SYStem,unknown (i.e random) parameters of interpretation OBjects,and those of picture BackGRounD.

The local criteria,thus,assign to each element of the processed picture a value of loss function averaged over random factors defining the given processed picture as well as over some subset of the processed pixels.The latter subset will be referred to as vicinity of the given pixel.

The notation of vicinity is widely used in picture processing.Most commonly,vicinity is understood in spatial sense,i.e. as a set of pixels geometrically surrounding the given one,and is defined by its area.Such vicinity of K pixels that are Spatially Nearest to the given pixel will be called KSN-vicinity.

The use of spatial vicinities in picture processing is due to the fact that picture and background details usually form spatial clusters.Another well-known property of images is the closeness of values of the pixels belonging to the same detail.It is only natural to add to the notion of spatial vicinity that of vicinity with respect to video signal value.Two kinds of such vicinities may be assumed - vicinity containing pixels differing from the given one at most by a predetermined value ε_V(Epsilon V) and that consisting of a given number K of pixels Nearest in Value to the given one.They will be referred to,respectively,as EV-vicinity and KNV-vicinity.

Another useful notion of vicinity may be related to the closeness of pixel positions in the so called variational row of pixels {V(R):V(R)<=V(R+1);R=1,2,...,N} ordered with respect to their increasing values.The ordinal number R(V) of an element in the variational row is called its rank,and the set of pixels whose rank differs from that of of the given pixel at most by some ε_R (Epsilon R)is called ER-vicinity.Formal description of EV-,KNV- and ER-vicinities may be conveniently carried out over an initial spatial vicinity:

$$EV(V(k,l))=\{V(R):V(k,l)-\varepsilon_V <= V(R) >= V(k,l)+\varepsilon_V\};$$

$$KNV(V(k,l))=\{V(R):\sum_{r=p}^{p+K-1} V(k,l)-V(R) = MIN\};$$

$$ER(V(R))=\{V(i):R(V(k,l))-\varepsilon_R <= i >= R(V(k,l))+\varepsilon_R\}.$$

This initial Spatial vicinity may be called S-vicinity,and EV-, KNV-,ER-vicinities may be,thus,regarded as the subset of the S-vicinity.

The variational row and ranks are closely related to the simplest statistical characteristic of pictures, histogram h(V) of distribution of pixel values:

$$H(V)=\sum_{m,n} \delta(V-V(m,n))$$

with δ (.) Kronecker's delta function because the following evident relation:

$$R(V)=\sum_{q=0}^{V} H(q)$$

holds for ranks and histogram. For present purpose, it is important that the introduced processing quality measure can be also expressed in terms of the histogram Hvcn(V) of elements belonging to the vicinity VCN defined by the locality function:

$$LOSSav(k,l)= AVimsys\{AVob\{AVbgrd\{\sum_{Vtr} Hvcn(Vtr)LOSS(Vtr(m,n);Vout(m,n)\}\}\}$$

with VCN={m,n:LOC(m,n;Vtr(k,l)=1}

Optimal processing is aimed at an estimate Vout(Vin(k,l);VCN) providing minimum of the averaged loss function LOSSav(k,l). Consequently, the corresponding algorithm will be optimal on the average. However, it is always desirable to determine an algorithm optimal for each particular picture which means that one must abandon averaging of the loss function over the background part of observed picture. Processing algorithms, if realizable, will be adaptive in this case to those random factors averaging over which is not included in the optimality criteion. Since the criterion defines the value of loss function within a local vicinity defined by the locality function, these algorithms will be locally adaptive. Further local criteria assume that all kinds of vicinities are determined in terms of initial spatial S-vicinity whose central element is the pixel to be estimated. This implies that the picture is to be processed in a fragment-wise manner, filter aperture that covers the fragment (S-vicinity) sliding over the processed picture.

It can be rather easily shown that all linear locally adaptive filters for picture processing follow from the above local criteria for KSN-vicinities and quadratic loss function while rank algorithms can be derived using EV-, KNV- and ER-vicinities.

References.

1. Bovic, A.C., Huang, T.S., Muson O.C. "A generalization of median filtering using linear combination of order statistics.". IEEE Trans. 1983. ASSP-31., p. 1342-1349
2. Hodgeson, R.M., Bailey, D.G.. Naylor M.J., Ng, A.L.M., McNeil S.T. "Properties, implementations and applications of rank filters", Image and vision computing, 1985, v.3, No1, p.3-14.
3. Jiang, S.S., Sawchuk, A. "Noise updating repeated filter and other adaptive noise smoothing filters using local image statistics" Applied Optics, 1986, v.25, No14, p.2326-2336.
4. Kim, V., Yaroslavsky, L.P. "Rank algorithms for picture processing Comp.Vis., Graph.Im.Proc., 1986, v.234-258.
5. Lee, J.S. "Digital image smoothing and sigma-filter", Comp.Vis. Graph., Im.Proc., 1983, v.24, p.255-269.
6. Vitkus, R.Yu., Yaroslavsky L.P. "Recursive algorithms for local adaptive linear filtration", In: Mathematical Research. Computer analysis of images and patterns/ed.by L.P. Yaroslavskii, A. Rosenfeld, W. Wilhelmi, bd.40. Academie Verlag, Berlin, 1987 p.34-39.
7. Yaroslavsky, L.P. "Digital picture processing. An Introduction", - Springer Verlag, New York, Heidelberg, Tokyo, 1985, pp.276.

SOME RESULTS ON GEOMETRICAL TRANSFORMATIONS IN THE DISCRETE PLANE

(Abstract)

by Albrecht Hübler
Friedrich-Schiller-Universität, Sektion Mathematik,
UHH, 17.OG, Jena, DDR-6900

This paper presents some further results in our investigations concerning the theoretical foundation of discrete geometry for image processing. The axiomatic theory of discrete geometry which was introduced in [3] has been developed on in [4,5]. It is essentially based upon the notion of translation. For a concrete model of this theory (In the planar case the well known grid point plane can be used as a standard model.) along with the set of translations such important geometrical notions like betweenness-relation, parallelity, orientation, halfplanes, and others are fixed. There were shown very interesting relations between this abtsract approach and the traditional digital geometry.

It is an important task, in addition to the class of translations to study further classes of geometrical transformations like rotations, and reflections, for example.
This can be done using different methods.

Let $\mathcal{P}$ be the set of points of the discrete plane. We define a mapping φ to be a *geometrical transformation* of the discrete plane, if φ is a bijection from $\mathcal{P}$ onto $\mathcal{P}$, and the betweenness-relation is invariant with respect to φ.
It can be shown [5], that the geometrical transformations are exactly those bijective mappings of the discrete plane which transform parallel straigt lines into parallel straight lines.
One way to introduce *rotations* (*reflections*) in the discrete plane [6] is to consider geometrical transformations with exactly one fixpoint (exactly one fixpoint straight line) fulfilling special invariance properties with respect to a fixed reference coordinate system. It turns out, that there are only very poor possible rotation groups, consisting of only 4 or 6 elements. These rotation groups are consistent (in the sense to build a motion group) with only 4 or 6 reflections, respectively.

These results are confirmed by results in [2] and in [1] where has been shown that any finite cyclic group of geometrical transformations consists of at most 6 elements, and that any finite group of geometrical transformations contains at most 12 elements.

Another way is to use metrics in the discrete plane in order to study the class of isometric geometrical mappings. This approach again leads to results consistent with those mentioned above. In [7] is shown that the "unit circles" generated by digital metrics which permit isometric geometrical transformations with exactly one fixpoint can have at most 6 "corner points". This implies that the corresponding isometric rotation group also can have at most 6 elements.

References:

[1] Baum. V.: Geometrische Transformationen in diskreten Räumen. Unveröffentl. Manuskript, Fr.-Schiller-Univ. Jena, Sektion Mathematik, 1989

[2] Bitter, M.: Some properties of the unimodular group in $\mathbb{R}^2$. In: A.Hübler, W.Nagel, B.D.Ripley, G.Werner (eds.): Geobild'89, Akademie-Verlag Berlin, Mathemat. Research, Vol.51 (1989), 37-41

[3] Hübler, A., R.Klette: Theoretische Grundlagen der digitalen Bildverarbeitung VII. Geometrie auf Nachbarschaftsstrukturen. Bild und Ton 40 (1987) 11, 229-235

[4] Hübler, A.: An axiomatic approach to discrete geometry and its relation to usual digital geometry for image processing. In: L.P.Yaroslavskij, A.Rosenfeld, W.Wilhelmi (eds.): Computer Analysis of Images and Patterns. Akademie-Verlag Berlin, Mathemat. Research, Vol. 40 (1987), 174-186

[5] Hübler, A.: Diskrete Geometrie für die digitale Bildverarbeitung. Diss.B, F.-Schiller-Univ. Jena, Sektion Mathematik, 1989

[6] Hübler, A.: Motions in the discrete plane. In: A.Hübler, W.Nagel, B.D.Ripley, G.Werner (eds.): Geobild'89, Mathemat. Research 51 (1989), Akademie-Verlag Berlin, 29-36

[7] Kotowski, O.: Der digitale metrische Raum $[\mathbb{Z}^2,\delta]$ und Eichfiguren. Diplomarbeit, F.-Schiller-Universität Jena, Sektion Mathematik, 1988

STRONGLY DILUTED NETWORKS WITH RETRIEVAL-DEPENDENT THRESHOLDS

H. Englisch[1], A. Engel[2], A. Schütte[3], M. Stcherbina[4]

1. Introduction

Though electronic elements work about a million times quicker than neurons standard programs on serial computers simulating human cognitive abilities perform much worser than our brain. Therefore new hardware solutions known under the name neural networks, parallel distributed processing or connectionism try mimick the architecture of the brain: Many simple, (mostly) uniform, highly interconnected processors (corresponding to neurons) operate parallely and exchange many informations. The information is stored robustly against partial perturbations in the whole system, also in the connections (corresponding to synapses) between the processors. Learning and self-organization of the network simplifies the programming drastically.

The best matching (minimum-distance classifier) is a typical task for the simplest networks: A memory which has stored p binary N-pixel patterns $\xi^\mu=(\xi_1^\mu,\dots,\xi_N^\mu)$, $\xi_i^\mu\in\{\pm 1\}$, $1\le\mu\le p$, may retrieve a specific pattern if at some time t=0 a perturbed version of that pattern $S(0)=(S_1(0),\dots,S_N(0))\in(\pm 1)^N$ is presented. In the easiest form the most likely original pattern is the one with maximal retrieval quality $m^\mu(0)$ with

$$m^\mu(t)=\sum_{i=1}^{N}\xi_i^\mu S_i(t)/N. \qquad (1)$$

The normalization 1/N guarantees $m^\mu(t)\in[-1,1]$. If the sign of the whole pattern does not play a role then $\max_\mu m^\mu(0)$ may be replaced by $\max_\mu |m^\mu(0)|$.

The grandmother (or Hamming) network (6) as the most suggestive solution for the determination of the pattern with maximal $|m^\mu(0)|$ has for $p\ll N$ the strongly diluted network by Derrida e.a. (4) as its descendent. (Confer (19) for comparison with classical search algorithms). In the parallely working version the diluted model obeys

$$S_i(t+1)=\mathrm{sgn}\Big(\sum_{j\neq i} c_{ij}J_{ij}S_j(t)+h_i\Big), \qquad (2)$$

where we included additionally pixel-dependent thresholds h_i. The sign function with sgn(0)=±1 is modified for the purpose of a bipolar output. The strength J_{ij} of the connection between pixel j and i is for

[1] NTZ Karl-Marx-Universität DDR-7010 Leipzig Karl-Marx-Platz 10
[2] Sekt.Physik Humboldt-Universität DDR-1040 Berlin Invalidenstr. 42
[3] Sekt.Elektronik Humbold-Universität DDR-1040 Berlin Invalidenstr. 110
[4] FTINT Academy of Sciences of Ucrainian SSR USSR-310164 Kharkow Lenin prosp. 47

random independent patterns with mean $<\xi_i^\mu>=0$ (corresponding to optimal coding) given by

$$J_{ij}=\sum_{\mu=1}^{p}\xi_i^\mu\xi_j^\mu/C, \qquad (3)$$

where C is a measure for the dilution: The random independent variables c_{ij} are 1 with probability C/N and 0 otherwise. The serial version to the parallel algorithm (2) reads

$$S_i(t+1/N)=\mathrm{sgn}(\sum_{j\neq i}c_{ij}J_{ij}S_j(t)+h_i), \quad S_j(t+1/N)=S_j(t) \quad \text{for } j\neq i, \qquad (4)$$

where i is a randomly or systematically chosen index. As time step we use 1/N in order to enable comparison with the parallel algorithm, where after one time unit also N new S-values are updated. The diluted model is studied for

$$1\ll C\ll N^{1/2t} \qquad (5)$$

in terms of $\alpha=p/C$. If $\vec{S}(0)$ is a perturbation of the νth pattern, independent of $\vec{\xi}^\mu(\mu\neq\nu)$, then the retrieval quality (we omit the index ν) fulfils for $h_i=0$

$$m(t+1)=f(m(t)),\ f(x)=\mathrm{erf}(x/\sqrt{2\alpha}),\ \mathrm{erf}(x)=2\int_0^x e^{-y^2}dy/\sqrt{\pi} \qquad (6)$$

in the parallel version and $dm(t)/dt=f(m(t))-m(t)$ in the serial one, resp. The limit retrieval quality m(oo) is decreasing with respect to α with $m(\infty)=1$ for $\alpha=0$, $1>m(\infty)>0$ for $0<\alpha<2/\pi=0,637$ and $m(0)=0$ for $\alpha>2/\pi$, cf. Figure 1. The convergence rate $c=\lim_{t\to\infty}-\ln(m(\infty)-m(t))/t$ is for the parallel version with $c=-\ln f'(m(\infty))$ larger than that for the serial one with $c=1-f'(m(\infty))$. Hence we consider in the following only the parallel version. Formula (6) is due to (5) in the non-diluted case C=N, the so-called Hopfield model [15], only valid for t=0, for $t\to\infty$ [2] or arbitrary $t\geq 1$ [1, 20] the retrieval quality was calculated only approximatively. (The dilute model can be solved rigorously also for C=O(1) [4], but the formulae are more complicate.) In order to increase the critical storage capacity in the Hopfield model either the J_{ij} are replaced by the matrix elements of a projection operator [7] or altered by time-consuming error-corrections [5] or the thresholds h_i are appropriately chosen [6]. The latter way should be gone in this paper. Other variants of the Hopfield model concern multipixel interactions for the storage of $O/N^k)$ patterns, $k>1$, [18] and memories with delay for the storage of pattern sequences [14].

From the point of view of the storage of real pictures it is important to weaken the assumption concerning independent patterns with mean zero: Modifications for the choice of J_{ij} in the case of hierarchically organized patterns [16], patterns with mean different from zero (i.e. with more white than black pixels) [2, 10] and spatially correlated patterns [9] has been proposed. Models with multistate pixels, representing different grey levels, are investigated in [13], whereas [12] is devoted to heteroassociative memories occuring e.g. in pattern classification.

2. The Input as Threshold

If $|\sum_{j\neq i} c_{ij}J_{ij}S_j(t)|$ is very small then the decision $S_i(t+1) = \text{sgn}(\sum_{j\neq i} c_{ij}J_{ij}S_j(t))$ is not reliable and it seems to be more favourable to put $S_i(t+1)=S_i(0)$, since $S_i(0)=\xi_i^\nu$ with probability $(1+m(0))/2$. This competition between information from the other pixel values and the stability of the pixel value considered can be taken into account by $h_i=h(t)S_i(0)$. For random independent patterns with mean zero formula (6) rewrites then to

$$m(t+1)=[(1+m(0))\text{erf}((m(t)+h(t))/\sqrt{2\alpha})+(1-m(0))\text{erf}((m(t)-h(t))/\sqrt{2\alpha})]/2. \tag{7}$$

The optimal h(t) is determined by

$$h_{opt}(t)=(\alpha/2m(t)) \;.\; \ln[(1+m(0))/(1-m(0))], \tag{8}$$

i.e. decreases with increasing m(t), but increases with increasing and m(0). As it should be expected with invreasing number of disturbing patterns or with improving input the input gets a higher weight. The retrieval quality m(t+1) is for the optimal h(t) a decreasing function of α with $m(t+1)>m(t)$. Thus if for all t the optimal parameter h(t) is chosen then $h_{opt}(t)$ decreases in time. If instead of the time dependent choice (8) the parameter h has to be fixed optimally the network with $h=h_{opt}(\infty)$ provides for $t\rightarrow\infty$ the same retrieval quality but the convergence is slower.
Nevertheless the convergence rates are the same. When iterating formula (8) numerically we did not find large differences for the number of necessary time steps between optimal fixed h and optimal decreasing h(t).

In Figure 1 the retrieval qualities $m(1)$ and $m(\infty)$ are shown for different initial qualities $m(0)$ and for $h_{opt}(t)$ as well as for $h=0$. The information gain $\eta(t)$ of $\vec{S}(t)$ in comparison to $\vec{S}(0)$ per pixel is according to the entropy formula

$$\eta(t)=[(1+m(t))\,ld((1+m)(t))/2)+(1-m(t))\,ld((1-m(t))/2) -(1+m(0))\,ld((1+m(0))/2)-(1-m(0))\,ld((1-m(0))/2)]/2. \qquad (9)$$

The ratio between the gains for all patterns $p\,.\,\eta(\infty)$ and the maximal information $C\,.\,ld(p+1).(1-C/2N)$ which can be stored in the C connections belonging to a pixel (every connection can possess $p+1$ different values; if $c_{ij}=c_{ji}=1$, then the coincidence of the two connection strengths $J_{ij}=J_{ji}$ is taken into account) is the information capacity

$$\gamma=\alpha\eta(\infty)/ld(p+1), \qquad (10)$$

tending for $p\to\infty$ to zero. With the help of non-linear synapses $J_{ij}\in\{\pm\sqrt{p/N}\}$ [11] γ can be increased by a factor $2\,.\,ld(p+1)/\pi$ also in diluted models [8]. Hence in Figure 2 we drew $\alpha\eta_{opt}(\infty)$ for h_{opt} as well as the difference $\alpha(\eta_{opt}(\infty)-\eta_0(\infty))$ to the corresponding quantity for $h=0$.

3. The Field Generated by the Input as Thres hold

In [6] the threshold $h_i=h^*(t).\sum_{j\neq i} c_{ij}J_{ij}S_j(0)$, which is hidden already in Kinzel's paper [17], was also investigated. For $t=0$ this threshold is without effect since it only multiplies the argument in (2) with the factor $(1+h^*(0))$. For $t>0$ we get

$$m(t+1)=\mathrm{erf}((m(t)+h^*(t)m(0))/\sqrt{2\alpha(1+2h^*(t)m(0)m(t)+h^*(t)^2}) \qquad (11)$$

and

$$h^*_{opt}(t)=m(0)(1-m(t)^2)/(m(t)(1-m(0)^2), \qquad (12)$$

whereas for $t=0$ the equation (6) is not altered. Since for $h^*(t)\to\infty$ the retrieval quality $m(t+1)$ tends to $m(1)$ for sufficiently large $h^*(t)$ $m(t+1)>m(1)$ is guaranteed. The optimal parameter $h^*_{opt}(t)$ decreases with increasing $m(t)$ and decreasing $m(0)$. In contrast to Section 2 the optimal parameter included in h_i depends on α only implicitly via $m(t)$. With respect to t $m(t)$ is increasing and $h^*_{opt}(t)$ is decreasing. For large α or large $m(0)$ the value $m(1)$ lies below $m(0)$ such that it can happen that $m(\infty)<m(0)$ though $m(\infty)>m(1)$. For $m(0)=0.8$ and $\alpha=0.45$ we get, e.g., $m(1)=0{,}767$ and $m(\infty)=0.815>m(0)$ in contrast to $\alpha=0.5$ with $m(1)=0.742$ and $m(\infty)=0.786<m(0)$.

As in the previous Section the network with fixed parameter $h^* = h^*_{opt}(\infty)$ yields for $t \to \infty$ the same retrieval quality as the optimal time-dependent network with $h^*(t)$ according to (12). Though the convergence rates coincide again the difference in the number of necessary iterations in order to get $(m(t)-m(t-1))$ less than some arbitrary bound (we have chosen 10^{-5}) should be demonstrated with an example: For $\alpha=0.5$ and $m(0)=0.4$ the final value 0.69814 is reached after 16 iterations with $h^*_{opt}(t)$ and 19 iterations with $h^* = h^*_{opt}(\infty)$, whereas after 36 iterations with $h^*=0$ the worser final value 0.61741 is received.
In Figure 3 $m(\infty)$ is drawn for the optimal h^* and $h^*=0$. Only for a small region in the parameter plane $(\alpha, m(0))$ around (0.7, 0.1) the choice for h_i of this Section yields remarkably better values than that of the previous Section. In the non-diluted model the approximations in [6] predicted smaller regions where the input as threshold performs better. In Figure 4 the difference between the values for $\alpha \cdot \eta_{opt}(\infty)$ calculated with $h_{opt}(t)$ and $h^*_{opt}(t)$ is shown.

Acknowledgement
We thank Mrs. T. Schröder for performing the numerical calculations.

4. Literature

(1) Amari, S., K. Maginu: Statistical neurodynamics of associative memory. Neural Networks 1 (1988) 1, 63-78.

(2) Amit, D.J., H. Gutfreund, H. Sompolinsky: Statistical mechanics of neural networks near saturation. Ann. Phys. (N. Y.) 173 (1987) 30-67.

(3) Amit, D.J., H. Gutfreund, H. Sompolinsky: Information storage in neural networks with low levels of activity. Phys. Rev. A 35 (1987) 2, 2293-2303.

(4) Derrida, B., E. Gardner, A. Zippelius: An exactly soluble asymmetric neural network model. Europhys. Lett. 4 (1987) 2, 167-173.

(5) Diederich, S., M. Opper: Learning of correlated patterns in spin-glass networks by local learning rules. Phys. Rev. Lett. 58 (1987) 9, 949-952.

(6) Englisch, H., A. Engel, A. Schütte: Improved retrieval in neural networks by means of input-dependent thresholds. To appear in J. New Generations Computer systems.

(7) Englisch, H., A. Engel, A. Schütte: Improved retrieval in neural networks with thresholds and ternary neurons. To appear in Proc. Conf.Dynamical Networks Eisenach 1988.

(8) Englisch, H., A. Engel, A. Schütte, M. Stcherbina: Improved retrieval in nets of formal neurons with thresholds and non-linear synapses. Submitted to Proc. Conf. Mathem. Psychology Berlin 1988.

(9) Englisch, H., M.V. Feigelman: In preparation.

(10) Englisch, H., J.L. van Hemmen: To be submitted to J. Stat. Phys.

(11) Englisch, H., M. Herrmann: Neural nets with non-linear synapses. To appear in Studia biophysica.

(12) Englisch, H., M. Herrmann: To be submitted to Proc. Evolution Theory and Combinatorial Optimization Eisenach 1989.

(13) Herrmann, M.: To be submitted to Proc. Neurocomputers and Attention Pushchino 1989.

(14) Herz, A., B. Sulzer, R. Kühn, J.L.van Hemmen: The Hebb rule: Storing static and dynamic objects in an associative neural network. Europhys. Lett. 7 (1988) 7, 663-669.

(15) Hopfield, J.J.: Neural networks and physical systems with emergent collective computational abilities. Proc. Natl. Acad. Sci. USA 79 (1982) 2554-2558.

(16) Иоффе, Л.Б.,М.В. Фейгельман: Асимметрия взаимодействий и иерархия образцов в моделях ассоциативной памяти. Письма в ЖЭТФ 44 /1986/3, 148-150.

(17) Kinzel, W.: Learning and pattern recognition in spin glass models. Z. Phys. B 60 (1985) 205-213; 62 (1986) 267.

(18) Lee, Y.C. e.a.: Machine learning using a higher order correlation network. Physica D 22 (1986) 276-306.

(19) Omohundro, S.M.: Efficient algorithms with neural network behavior. Complex Syst. 1 (1987) 2, 273-347.

(20) Zagrebnov, V.A., A.S. Chvyrov: Little-Hopfield model: recursion relation for retrieval pattern errors. Preprint JINR Dubna, USSR.

Figure Captions

1. The retrieval quality $m(1)$ (Fig. 1A) and $m(\infty)$ (Fig. 1B) for $h=h_{opt}$ (full line) and $h=0$ (broken line) with $m(0)=0,2$ (a), $m(0)=0.5$ (b) and $m(0)=0.8$ (c). The parallels mark the initial retrieval quality $m(0)$. In Fig. 1B $m(\infty)$ does not depend on $m(0)$ for $h=0$.

2. The total information gain per pixel $\alpha . \eta_{opt}(\infty)$ for h_{opt} (full line) and the difference $\alpha .(\eta_{opt}(\infty)-\eta_0(\infty))$ broken line) with $m(0)$ as above. The cusp in the broken line at $2/\pi$ reflects the cusp of $m(\infty)$ for $h=0$ (cf. Fig. 1B).

3. The retrieval quality $m(\infty)$ for $h^*=h^*_{opt}$ (full line) and $h^*=0$ (broken line) with $m(0)$ as in Fig. 1.

4. The total information gain $\alpha . \eta^*_{opt}(\infty)$ for $h^*=h^*_{opt}$ (full line) and the difference $\alpha . (\eta_{opt}(\infty)-\eta^*_{opt}(\infty))$ (broken line) with $m(0)=0.05$ (a), $m(0)=0.125$ (b) and $m(0)=0.2$ (c).

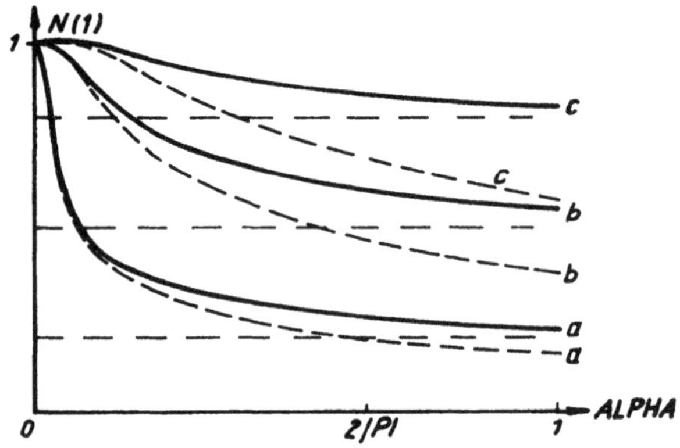

Fig. 1a

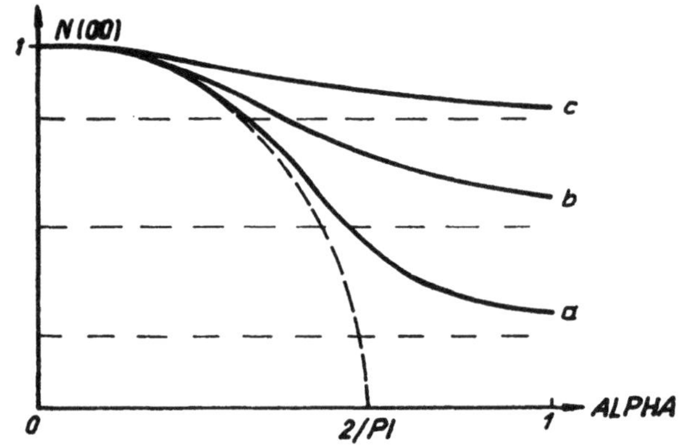

Fig. 1b

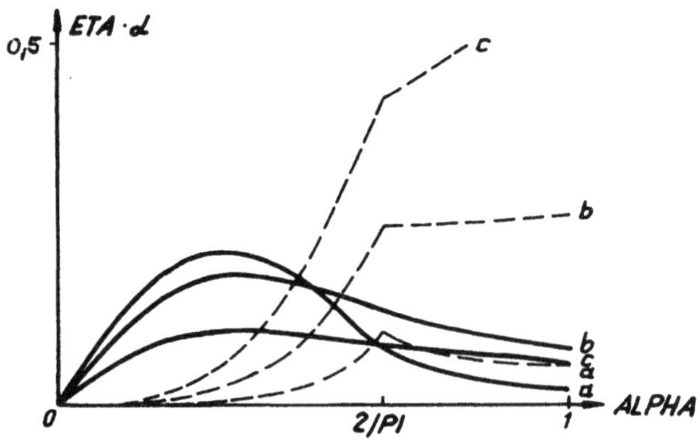

Fig. 2

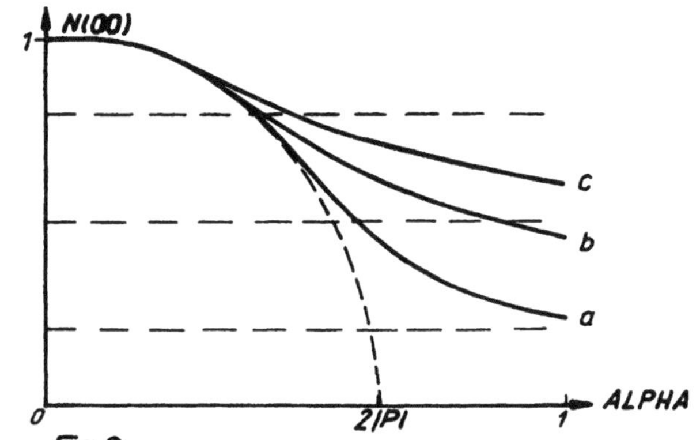

Fig. 3

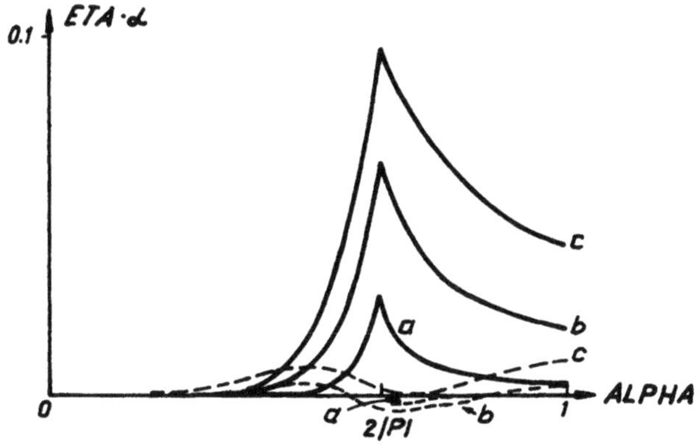

Fig. 4

MODELLING OF 3D-OBJECTS FROM MONOCULAR TV-IMAGE SEQUENCES

C.-E. Liedtke, F. Kappei 1)

Abstract

An approach is presented, where a 3D scene consisting of 3D moving objects in a 3D space is modelled from a sequence of monocular TV images. The parameters of a parametric model are extracted automatically during an analysis phase. In a later synthesis phase the 3D-model is used to reproduce the original input image sequence from the camera. The system is primarily developed for future image transmission systems working at very low data rates, as they are needed for TV conferences. Other applications comprise rate conversion, extraction of stereo sequences from monocular TV sequences and general applications in Computer Graphics.

Introduction

Luminance changes in image sequences can frequently be interpreted as caused by movements of 3D objects in a 3D space. This appears to be obvious in those situations, where the image sequences have been obtained from a TV camera observing the physical 3D world. In picture coding applications, modelling of planar objects in connection with 2D movements are often used [1] . The disadvantages of the 2D models become apparent in the interpretation of 3D effects like covering and uncovering of objects, rotation, zooming and scale variations, when the objects move towards or away from the camera. In order to obtain a higher accuracy in modelling it becomes necessary to extend the models to 3D objects and movements in 3D space.

Dreschler [2] has proposed an approach for the assessment of a 3D object description based on the use of prominent points. The disadvantage of this approach for "natural scenes" is, that prominent points are usually not available to the extent which is necessary. A different approach has been proposed by Waxman [3] , who uses optical flow for modelling. Here difficulties arise at the edges of moving objects. It appears further from his publications that there are difficulties in the modelling of arbitrary surfaces from the optical flow.

In order to overcome these difficulties we use a model based image analysis procedure which is based on Analysis-by-Synthesis-Techniques. Since the aim of the project is to explain luminance changes in image sequences from a camera in a transmission system in order to reconstruct the image sequence from the model at a receiver we characterize the model by the

---

1) Prof. Dr.-Ing. C.-E. Liedtke, Dr.-Ing. F. Kappei, Universität Hannover, Institut für Theoretische Nachrichtentechnik und Informationsverarbeitung, Appelstr. 9A, 3000 Hannover 1, FRG

term "projection model". The projection model is close to but not necessarily identical to the exact 3D description of objects in the physical world.

Projection Model

Image sequences which have been recorded from a standard TV camera are explained as 3D objects moving in three dimensions, which have been illuminated from an illumination source and where the reflected light has been projected onto a camera target. Therefore the projection model consists of three components: (1) the camera model, (2) the illumination model and (3) the object model. Since an automated modelling in all generality seems presently not to be possible, the following restrictions for the modelling have been required: (1) The illumination is diffuse, (2) the objects are "quasi-rigid", and (3) the objects are opaque and have a matte surface.

The camera model describes the projection of a 3D object onto the 2D image plane. The perspective projection is used. Parameters are the focal length and the number of rows and columns of the image plane. The illumination model describes the illumination of the whole scene. The intensity of the diffuse light source serves as parameter.

The object model is a complete description of the visible part of the moving objects. Parameters are the motion parameters, shape parameters and surface properties of the individual objects: The motion parameters require for each object six parameters to describe translation and rotation in space. The shape of an object is described by a closed mesh of surface triangles. The vertices of the triangles are described by so called control points. The positions of these control points in space constitute the shape parameters. The surface properties of the objects are described by the gray levels on the object surface. The surface properties are once stored and then used during image reconstruction by reprojection of the corresponding part of the stored original image onto the object surface.

Model Adaptation

The parameters of the parametric model are determined fully automatically by an Analysis-by-Synthesis-Technique. For each new incoming image from the camera an image is predicted by reconstruction from the model. Differences between the incoming image and the predicted image are used to adapt the model. Of special importance is the adaptation of the object model. Shape and movement parameters are obtained in an iterative procedure. The shape adaptation includes a global adaptation, where groups of control points are adjusted simultaneously, a local adaptation where single control points are adjusted in the direction of the viewing-line, and movements of control points perpendicular to the viewing-line in order to model elastic surface properties of the objects. The viewing line is the line connecting the focal point of the camera and the point P under consideration on the surface of the 3D object model.

The camera model is shown in Fig. 1. $\mathbf{P} = (P_x, P_y, P_z)^T$ is a point on the surface of the 3D objects model. It has the gray-level value I. $\mathbf{B} = (B_x, B_y)^T$ represents the projection of P onto the image plane under consideration of the focal length F. After the spatial movement of the object has been registered by observation of a consecutive image sample the motion parameters are calculated from the difference between the expected gray level I and the recorded true gray level I(B") observed in the image plane. B" in the image plance corresponds to the predicted spatial movement of P into P". The linearized motion equations can for any visible point P" be formulated as

$$
\begin{aligned}
I\,(B'') - I = {} & \\
& F*G_x/P_z''*T_x' \\
& + F*G_y/P_z''*T_y' \\
& - F*(P_x''*G_x+P_y''*G_y)/P_z''^2*T_z' \\
& - F*(P_x''*G_x*(P_y''-C_y)+P_y''*G_y*(P_y''-C_y)+P_z*G_y*(P_z''-C_z))/P_z''^2*R_x' \\
& - F*(P_y''*G_y*(P_x''-C_x)+P_x''*G_x*(P_x''-C_x)+P_z*G_x*(P_z''-C_z))/P_z''^2*R_y' \\
& - F*(G_x*(P_y''-C_y)-G_y*(P_x''-C_x))/P_z''*R_z'
\end{aligned}
\tag{1}
$$

with	$G = (G_x,G_y)^T$		Gradient of a surface element
	$C = (C_x,C_y, C_z)^T$		Center of gravity
	$T' = (T_x',T_y',T_z')^T$		Translation vector
	$R' = (R_x',R_y',R_z')^T$		Rotation vector

For a selected number of points on the surface of the model Eqs. (1) are solved for the unknown vectors T' and R'. Since the linearization of the motion equations requires an iterative procedure for the calculation of the motion parameters, new estimates for the true translation and rotation vectors are

$$R_{new} = R_{old} + R' \tag{2a}$$

$$T_{new} = [R'] *(T_{old} - C)+ T' \tag{2b}$$

With a similar approach the shape of the 3D object is calculated. In this case the "movement" of a surface point into a position of improved shape representation is described by

$$P' = [R] *(P + V * S) + T \tag{3}$$

where in addition to translation T and rotation [R] the "movement" is described by a shape parameter S along the viewing line V.

The shape adaptation process has in more detail been described by the authors elsewhere [4, 5] .

The result of the shape modelling from just two incoming images according to Fig. 2a and Fig. 2b is illustrated in Fig. 2c and Fig. 2d. The latter two illustrations show the spatial mesh

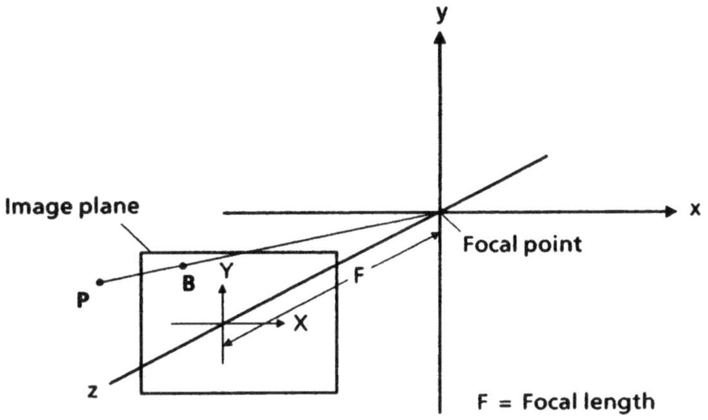

Fig. 1 Camera model

of triangles, which forms the surface, under two different angles. For the reconstruction of images, the surface triangles need to be complemented by their surface properties, which have been obtained and stored once from the incoming image sequence. Employing these surface

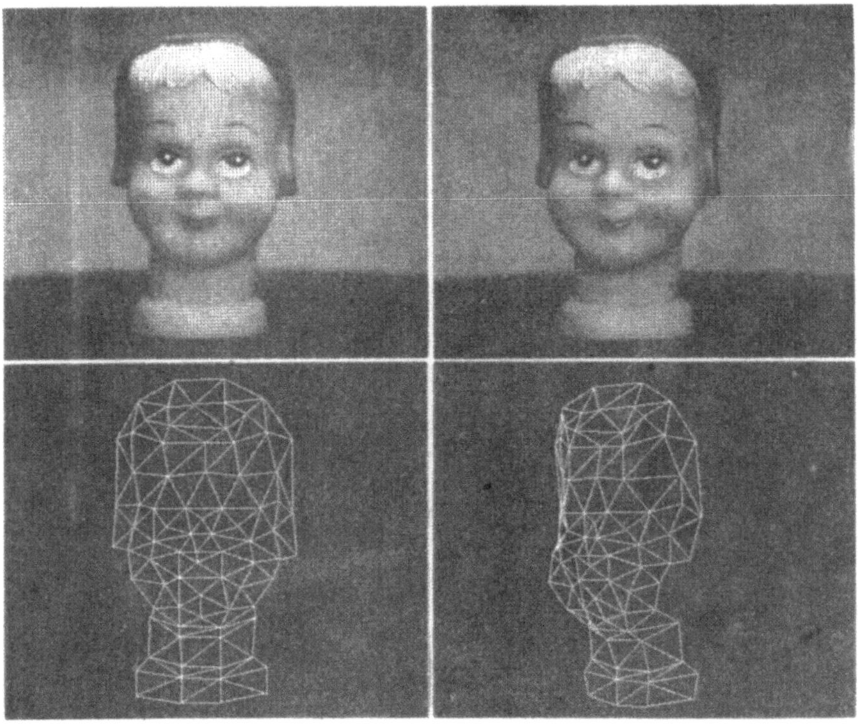

Fig. 2 Shape modelling: Original images at 0° (a) and 5° (b), derived wire frame models at 0° (c) and 30° (d)

a	b
c	d

properties, reconstructions at different spatial positions become possible as is shown in Fig. 3. Fig. 3d appears not to be subjectively satisfactory because the reconstruction of the head under 20° requires the availability of parts of the head, which were not visible in the two original images of Fig. 3a (0°) and Fig. 3b (5°) from which the model had been derived.

Results

The method has been applied successfully to the modelling of the complete surfaces of objects. The complete reconstruction of the puppet head from Fig. 2 and Fig. 3 became possible from a sequence of 9 camera images which have been recorded while turning the puppet head on a turntable by 80°. The method has further been tested on internationally agreed test sequences for video telephone research like the sequence "Miss America".

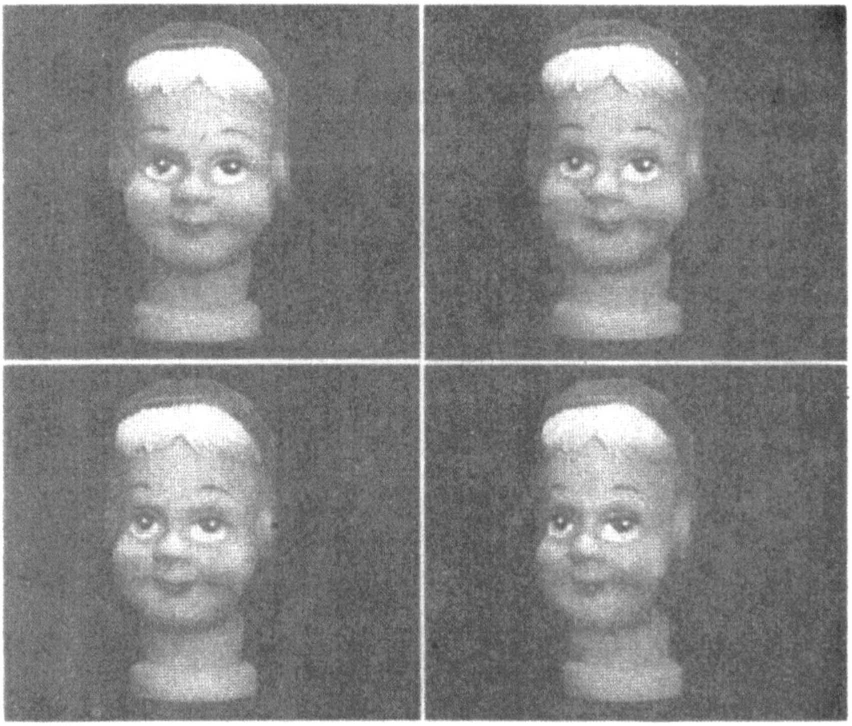

Fig. 3 Image reconstruction at angle 0° (a), 5° (b), 10° (c), 20° (d)

a	b
c	d

Literature

[1] H.G. Musmann, P. Pirsch and H.-J. Grallert: "Advances in Picture Coding", Proc. IEEE, Vol. 73, pp. 523 - 530, April 1985

[2] L. Dreschler: "Ermittlung markanter Punkte auf den Bildern bewegter Objekte und Berechnung einer 3D-Beschreibung mit dieser Grundlage", Dissertation, Universität Hamburg, 1981

[3] Allen M. Waxman: "Closed Form Solutions to Image Flow Equations for 3-D Structure and Motion", Computer Sciences Technical Report Series, University of Maryland, 1986

[4] Kappei, Liedtke: "Modelling of a Natural 3-D Scene Consisting of Moving Objects from a Sequence of Monocular TV Images", SPIE, Cannes 1987

[5] Kappei: "Modellierung und Rekonstruktion bewegter dreidimensionaler Objekte in einer Fernsehbildfolge", Ph.D-Thesis, Universität Hannover, 1988

COMPUTING FRACTAL DIMENSIONS OF IMAGE SEGMENTS

REINER CREUTZBURG

Academy of Sciences of the G.D.R.
Central Institute of Cybernetics
and Information Processes ·
Image Processing and Computer Graphics
Kurstr. 33, P.O.Box 1298
DDR - 1086 Berlin

EVGENI IVANOV

University of Technology
Dept. of Information Techniques
Mommsenstr. 13
DDR - 8027 Dresden

1. INTRODUCTION

Many objects in images of natural scenes are so complex and irregular, that describing them by the familiar models of classical geometry is insufficient.

The concept of the fractal dimension can be useful in the analysis and classification of shape and texture of images and image segments [Man82], [Flo79]. This concept appears to us very important, because fractals are of close similarity to the natural surfaces and shapes and can be even interpreted as the result of many basic physical processes. A. Pentland noticed that "the fractal model of imaged 3-D surfaces furnishes an accurate description of most textured and shaded image regions " [Pen84].

These properties of the fractal model have influenced its wide spread in the recent years in topology [Cla86], sedimentology [OrW83] and particle morphology [Kay86], description of natural scenes [Pen84], image segmentation [Rig88], compression, coding and computer graphics [PeS88]. It has become a powerful tool in the image processing and we are shure, that its importance will increase.

The aim of this paper is to

- develop a new fast algorithm for calculating the fractal dimension of an image surface,
- analyze automatically global and local image structure characterizing them by their total and fine fractal dimensions and estimating the boundary between image texture and image structure,
- implement the algorithms on a PC and on a superminicomputer.

2. THE FRACTAL DIMENSION

A fractal is defined as a set, for which its Hausdorff-Besicovich dimension strictly exceeds its topological dimension (for example 1, 2 or 3). The main characteristic of a fractal is its fractional (fractal) dimension, where the fractal has its name from. (The fractal dimension has not to be an integer - "fractus" = non integer.)

The fractal dimension can be computed by subtracting a fractional quantity from the classical (topological) dimension D_T. This fractional quantity will be denoted by β (fig.1).

$$D = D_T - \beta .$$

If an object can be described by its fractal dimension, the graph of the logarithm of the estimated perimeter or surface area against the logarithm of the resolution, at which the estimation has been computed, tends towards a straight line data relationship, known as the RICHARDSON-MANDELBROT-plot. The calculated perimeter or surface area obviously increases when the resolution decreases, therefore the slope of the relationship is always negative. It tends towards zero for any regular Euclidean object and becomes non-zero, when irregular Euclidean outlook is viewed. The slope of the relationship can be computed by a correlation function or by a linear regression and is the above mentioned fractional quantity β . This estimation bases sometimes on a fit through some few points, which often deviate from a straight line. Obviously, the fractal dimension defined this way supplies only global information about the entire outlook of the analyzed images and is sufficient only for an ideal, fully self-similar fractal object. That's why we call it total fractal dimension and will denote it further by D_{tf}. Nevertheless, the total fractional dimension provides useful characteristics of the analyzed objects and is absolute necessary for classification problems. We consider and use it as the base of any further computation and analysis.

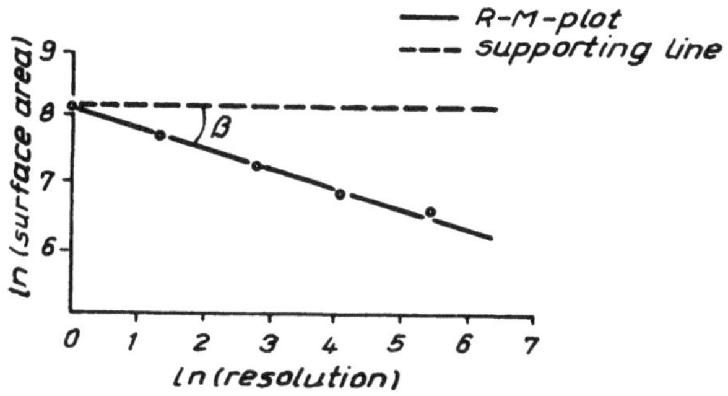

Fig. 1: Determination of the fractal dimension

3. THE FINE FRACTAL DIMENSIONS

The recent results of the research works in the field of the fractal model and the fractal geometry point out that most of the natural objects are not ideal but semi-fractal. It means, there is a change in the original model of self-similarity, as far as the deviation from the straight line RICHARDSON-Mandelbrot-plot could be interpreted this way. The expression of this phenomenon is a break in the RICHARDSON-MANDELBROT-plot and already Flook in 1979 [Flo79] spoke about an empirical evidence, that often two linear regions appear in the plot as two separate fractional elements.

The first element, observed at smaller dilåtation of the resolution, characterizes obviously the finest edge or surface effects of the outline of the analyzed image. It is called by Kaye and Flook the textural fractal element and its quantitative measure is the textural fractal dimension, which we denote by D_1 with

$$D_1 = D_T - \beta_1$$

(see fig. 2). The textural fractal dimension provides therefore information about the irregular Euclidean object at a very low level of resolution and consequently of abstraction.

The second element, beginning after the break point in the slope, was refered to by Flook as the structural fractal element, because it describes the object at a higher resolution level, it means, it describes its coarse nature, its structure (see fig. 2). This element refers to the macroscale outlook of the image. The fractional dimension, related to this element, is called the structural fractal dimension D_2 with

$$D_2 = D_T - \beta_2 .$$

Quantitatively evaluated, the absolute value of the sum of the fractional quantities and of D_1 and D_2, respectively, is always bigger than (or in the boundary case equal to) the absolute value of the fractional quantity of the total fractal dimension D_{tf}, whereby D_{tf} is either smaller or bigger than the textural or the structural fractal dimensions, and the three plots could be thought of as of the sides of a triangle. Besides, D_1 and D_2 possess the same qualities as already pointed out by D (previous paragraph), and can be defined and calculated in the same way, applying the same algorithm but for different resolutions.

Although the estimation of these two new fractal dimensions brings our classification problem a step further, it is still not sufficient. Because of the above mentioned qualities it is necessary to determine which of these two dimensions is dominating. Saying it in a different way, to decide, whether the two-element fractal model (and so the RICHARDSON-MANDELBROT-plot) is convex, concave or even a straight line (fig. 3). Here we calculate the quotient Δ of the two fine fractal dimensions and use it for this purpose. This quotient is defined by

$$\Delta = D_1/D_2 .$$

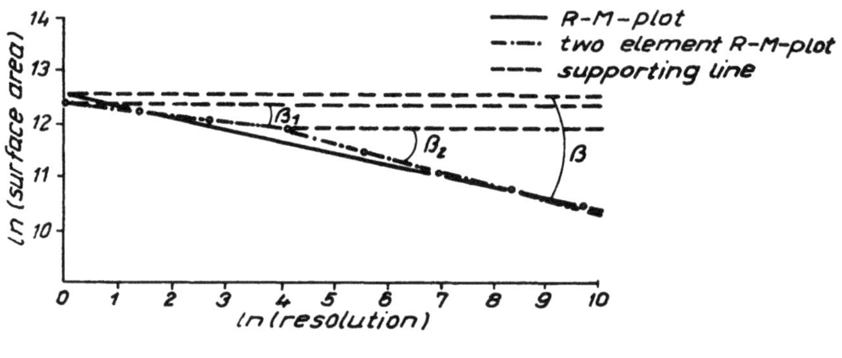

Fig. 2: Determination of the fine fractal dimensions

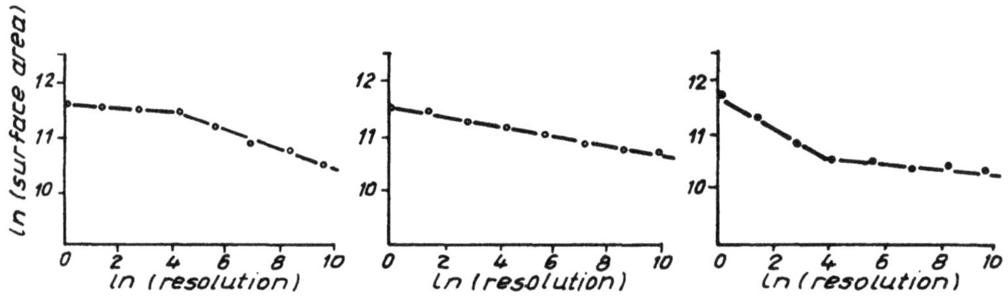

a) b) c)

Fig. 3: Different fractal models (RICHARDSON-MANDELBROT-plots)
a) convex; b) linear; c) concave.

According to the value of Δ one can distinguish among three classes of fractal objects:

a) Δ strictly greater than one: texture is dominant (texture of the analyzed image is more irregular than its structure),

b) Δ equal to one, or close to one: texture and structure are equally dominant (analyzed image is fully self-similar fractal),

c) Δ strictly less than one: structure is dominant (texture of the analyzed image is smoother and less irregular than its structure).

If we find, that the analyzed image fits closely to the two-element fractal model, it is also necessary to define the boundary of the two fine fractal models. Our experiments showed, that the break point abscissa coordinate X_b is well-suited for a boundary definition (fig. 4). The evaluation of the last definition sets some special requirements on the applied algorithm, i.e. on the computation of this value (see chapter 4).

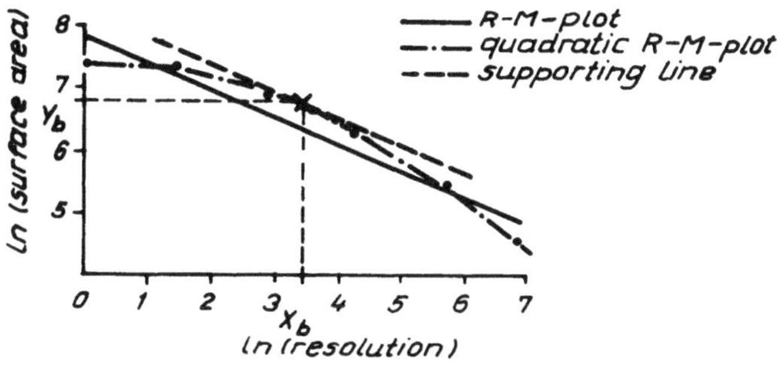

X_b - break point abscissa
Y_b - break point ordinate

Fig. 4: Break point determination in the RICHARDSON-MANDELBROT-plot

4. ALGORITHM FOR COMPUTING THE FRACTIONAL DIMENSION OF GREY-LEVEL IMAGE SURFACES

In this chapter we propose a new fast algorithm for calculation of fractal dimensions of grey-level image segments in a 3-dimensional space (i.e. topographic surfaces, grey-level images and others).

The method uses a 3-dimensional equivalent of the "walking dividers" method in two dimensions. Consider a digitized image, where the appropriate values (of elevation or grey-tone) are ordered at the intersections on a regular square grid of uniform spacing in a coordinate system (we used Cartesian coordinates, but it is not absolute necessary).

Given four pixels A, B, C, D, defining a square window, gliding over the grid without overcovering. We call them corners. Let us denote the corresponding values by a, b, c, d, and define a center of the square E and attach to it an interpolated value e, according to the formula

$$e = (a + b + c + d)/4.$$

So we can divide the square into four triangles connecting the corners with the center point - ABE, BCE, CDE, DAE. All triangles with their corner values, which can be understood as elevations define triangular prisms, which tops are part from the original grey-level image, whose surface area we want to calculate (fig. 5). The main idea of the algorithm is that we approximate the analyzed image by planar triangles, whose surface area is easy to compute, at different levels of resolution.

We calculate the surface area of the tops of the prisms by means of the vector algebra (vector multiplication). Each side of a triangle defines a vector, if we attach a direction to it. Therefore it is only necessary to find out the vector coordinates, which is almost no problem, if we consider, that we want to compute a vector product and so we can change the origin of the coordinate system for every new triangle. We use the computation of the determinants of the coordinates to calculate the vector products.

Given two vectors $\mathbf{a}$ and $\mathbf{b}$, with Cartesian coordinates

$$\mathbf{a} = (x_a, y_a, z_a) \quad \text{and} \quad \mathbf{b} = (x_b, y_b, z_b).$$

The vector product

$$\underline{a} \times \underline{b} = \left(\begin{vmatrix} y_a & z_a \\ y_b & z_b \end{vmatrix}, \begin{vmatrix} x_a & z_a \\ x_b & z_b \end{vmatrix}, \begin{vmatrix} x_a & y_a \\ x_b & y_b \end{vmatrix} \right) ,$$

and the surface area of the triangle spanned by these two vectors is with already computed determinants

$$S = 1/2 * \sqrt{(y_a z_b - y_b z_a)^2 + (x_a z_b - x_b z_a)^2 + (x_a y_b - x_b y_a)^2}$$

So we need only to calculate the vector coordinates, whereby the origin of the coordinate system is wandering clockwise on the corners of the window, starting at point A and is kept constant for each triangle. The total surface area of the top of a grid cell is computed by adding of the top surface areas of the triangular prisms, defined in this cell:

$$S_{ABCD} = S_{ABC} + S_{BCE} + S_{CDE} + S_{DAE} ,$$

and the total top surface area of the grey-level image is the aggregated surface area of all grid cells (square windows). This operation can be repeated for dilatating cell sizes, and as already pointed out, the total top surface area will decrease.

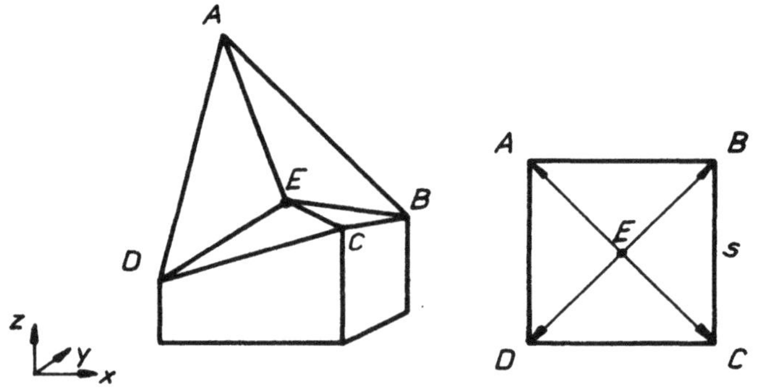

triangular prism representation top view

Fig. 5 : Grid cell with a sidelength equal to s

We used cell sides equal to powers of two, starting by one (it means involving every pixel in the image) and ending by the largest power of two, smaller at least by one from the image size. This proved to be very useful, because a power of two (i.e. 256 x 256, 512 x 512 or even bigger) is a standard image size and defining the upper boundary this way, we prevent some undesired influences of the image surroundings. It also provides an uniform spread of the resolution in the logarithmic scale, where the fractal dimension is computed. Because of the applied algorithm we use only 72 simple floating point operations (addition, subtraction and multiplication) and only 4 square root operations for calculating the top surface area of a grid cell. Detailed comparison of similar algorithms is listed in table 1.

It has to be noted, that the square root and the absolute value floating point computations are standard subroutine calls and are therefore very time consuming.

The total fractal dimension is calculated according to the above explained formula $D_{tf} = D_T - \beta$, where D_T equals 2 as we analyze two-dimensional images. The fractional quantity β is the slope of the regression line, approximated through the calculated values in the double-logarithmic scale by a correlation function. The correlation coefficient, denoted by r, is interpreted as a measure for the deviation from the ideal fractal model.

OPERATION	ALGORITHM	
	Clarke in [Cla86]	new method
addition	29	21
subtraction	28	24
multiplication	36	27
division	1	--
square root	12	4
absolute value	4	--
total	110	76

Table 1: Number of operations, necessary to approximate the top surface area of a grid cell with different algorithms

After computing the total fractal dimension, the estimated values for each different spacing are approximated to a quadratic polynomial.

The break point in the already estimated RICHARDSON-MANDELBROT-plot is defined as the point of the parable, described by the quadratic polynomial, at which the parable and the straight line, obtained by the estimation of the total fractal dimension, possess the same slope β (it means their first derivatives are equal) (see fig. 4). Therefore the formulas for calculation of the break point coordinates are

$$X_b = 1/2 * (\beta - a_1) / a_2 ,$$

$$Y_b = a_0 + a_1 X_b + a_2 X_b^2 .$$

This definition was chosen, because of the already introduced properties of D_{tf}, D_1, D_2 (see chapts. 2 and 3).

Once found, the break point abscissa X_b is used to divide the fractal model into two parts - textural (on the left side of the break point) and structural (on the right side of the break point) fine fractal model (see chapter 3).

Then, the textural and the structural fractal dimensions are estimated by correlation functions, whereby the correlation coefficients r_1 and r_2 can be interpreted as a measure for the deviation from the particular fractal model.

Finally, the quotient $\Delta = D_1/D_2$ is computed.

It has to be noted, that the decision, if the one- or the two-element fractal model fits better the analyzed image, has to be made after computing all parameters D_{tf}, D_1, D_2, r, r_1 and r_2. A break point exists only if

i) Δ differs noticeable from one;

ii) r_1 and r_2 are far closer to one than r;

iii) the break point abscissa X_b can be attached neither to the first nor to the last resolution step (in such cases the estimated values at these resolution steps are treated as computational errors).

It has also to be pointed out, that the total fractal dimension D_{tf} is computed only if there are at least four observation points (four different resolution steps), and the fine fractal dimensions D_1 and D_2 - only if there are at least five observation points. These decisions are necessary only in the sense of more statistical security.

So the smallest images, which can be analyzed with the proposed algorithm have to be at least 9 x 9, if we want to apply the one-element fractal model, and at least 33 x 33, if we want to apply the two-element fractal model.

Obviously the statistical security is higher, if we increase the image size. But it looks out as an optimization problem, as far as the size dilatation will increase the elapsed computational time, necessary for the analysis and bring more complexity in the image and this way -

problems by the application of the fractal model in general (mainly by classification problems, because of the necessity of uniqueness of the made decision), so this problem has to be thought over carefully and will be the topic of our further investigations.

A computer program in C was implemented on an IBM PC-AT under MS/DOS and on a 32-bit superminicomputer SM 52 running under VMS. The program is able to read digital image matrices and computes the total and fine fractal dimensions, the quotient of the fine fractal dimensions and the break point coordinates of an image surface or image segment. Because of the used algorithm, it is much faster than comparative ones. Comparison between the elapsed time using similar algorithms is shown in table 2.

IMAGE SIZE	FRACTAL/[Cla86]		new method	
	IBM PC-AT	SM 52	IBM PC-AT	SM 52
17	8	2	5	1
33	31	4	17	2

Table 2: Comparison between the elapsed time in seconds for computing the fractal dimensions with different algorithms

The efficiency of the proposed algorithm was proved by the results of the various application examples.

REFERENCES

[Cla86] Clarke, K. C.: Computation of the fractal dimension of topographic surfaces using the triangular prism surface area method. Computers & Geosciences 12 (1986) No. 5, pp. 713-722.

[Flo79] Flook, A. G.: The characterization of textured and structured particle profiles by the automated measurement of their fractal dimension. Partikel Technologie Nuernberg 24-26.IX.1979, pp. 591-600.

[Kay86] Kaye, B. H.: Fractal dimension and signature wave form characterization of fine particle shape. American Laboratory Apr. 1986, pp. 55-63.

[Man82] Mandelbrot, B. B.: The Fractal Geometry of Nature. W. H. Freeman: San Francisco 1982.

[OrW83] Orford, J. D., Whalley, W. B.: The use of the fractal dimension to characterize irregular-shaped particles. Sedimentology 30 (1983), pp. 655-668.

[Pen84] Pentland, A. P.: Fractal-based description of natural scenes. IEEE Transactions on Pattern Analysis and Machine Intelligence PAMI 6 (1984), No. 6, pp. 661-674.

[PeS88] Peitgen, H.-O., Saupe, D. (Eds.): The Science of Fractal Images. Springer: Berlin 1988.

[Rig88] Rigaut, J. P.: Automated image segmentation by mathematical morphology and fractal geometry. Journal of Microscopy 150, Pt. 1 (1988), pp. 21-30.

TOWARD LINE DETECTION BY CLUSTER ANALYSIS

M.J. Silbermann *   J. Sklansky *

Abstract

We propose a new method to detect straight lines embedded in noise, and to determine their parameters (ρ, θ) with high accuracy. This method transforms selected pairs of points into the Hough domain where cluster analysis is applied to detect subsets of strongly collinear points in the image.

1 Introduction

The detection of straight line segments in a noisy image is a problem that occurs in numerous image analysis applications. Several methods are currently used to solve this problem. One of the most popular approaches especially for high levels of noise is known as the Hough transform [1], [2], [3]. Duda and Hart replaced the problem of finding collinear points in an image by the mathematically equivalent problem of finding concurrent sinusoidal curves in a 2-dimensional parameter space. The interest generated in the image processing community by the Hough transform is indicated by the abundance of literature in the field (a recent and thorough survey is given in [4]). Unfortunately, although this technique is qualitatively well understood, its quantitative performance (accuracy in detection and localization) is not adequately documented.

Our major interest here is the accuracy of the estimated positions of the detected lines. Generally it is the quantization of the parameter space, as well as the speading of the peaks in that space that determines the accuracy. In [5], H. Maitre shows that the optimal quantization is obtained if the integral of the count probability for a uniform random noise, over any cell in the parameter space, is constant. In [6], T. Van Veen and F. Groen discuss the relationship between $\triangle\rho$ and $\triangle\theta$, where ρ and θ are the parameters and $\triangle\rho$ and $\triangle\theta$ are the resolution for each parameter. They show that an optimal value of $\triangle\rho$ as a function of $\triangle\theta$ can be derived if the maximum length of the line is known. In [7], W. Niblack and D. Petkovic propose a technique that uses interpolation to determine the peak value at a sub-cell resolution. In that case the accuracy is finer than the quantization of the parameter space.

In this paper we propose a new technique that is independant of quantization in the parameter space and that achieves high accuracy in the localization of the lines. This method is based on Hough's original idea of using a parameter space in which each object to be detected (for example lines, cercles, parabolas) is represented by one single point. Our method differs from Hough's in that the points in parameter space are generated from information in a neighborhood of each processed image pixel, while in the traditional Hough transform no context information is used.

2 The Hough Transform

The Hough transform maps straight lines from the image space into a parameter space such that each point in the parameter space represents a single straight line in the image space. (For convenience, we will use hereafter "line" to denote "straight line").

The most common representation of a line is given in terms of its slope m and its y-intercept b :

$$y = m\,x + b \tag{1}$$

*University of California, Irvine, Department of Electrical Engineering, Irvine, CA 92717

This representation, however, shows a singularity for lines that are nearly vertical, i.e. $m \to \infty$, $b \to \pm\infty$. Duda and Hart [1] suggested the use of the normal representation of a line which is given by its orientation, θ, and its normal distance to the origin, ρ (see Figure 1a).

$$\rho = x\, cos\theta + y\, sin\theta \tag{2}$$

With this representation every line in image space corresponds to a point in parameter space.

This transformation has been used in the last decade for line detection in digital binary valued images in the following way. Each discrete data point in the x-y domain (image space) is transformed into a curve in the ρ-θ domain (parameter space), which is quantized into cells. An accumulator is assigned to each cell. At all the cells associated with the computed values ρ and θ, the accumulators are incremented by 1. After all the data points are transformed, the parameter space is examined. Large accumulator counts correspond to collinear data points that may be fitted by a straight line. Small counts generally indicate either isolated data points or noise points that can be deleted.

One of the attraction of the Hough transform is its insensivity to gaps (partial occlusions).

3 Presentation of the new method

We propose a new method to estimate the parameters of straight lines embedded in noise with high accuracy. In the Hough transform each image pixel (i, j) generates a curve C_{ij} in the parameter space. The technique we present here uses local information around each image pixel to generate a small number of points in the parameter space. Those points lie on the curve C_{ij} and they contain all the useful information. In other words the new method considers only a small subset of possible lines. The idea is that each image pixel (i, j) is associated with a set of its neighbors $\{(p, q)\}$. For every pair of pixels $((i, j), (p, q))$ in image space, we compute the point (ρ, θ) in parameter space which represents the line passing through (i, j) and (p, q). A cluster analysis is then applied to the points (ρ, θ) to find sets of strongly collinear points. In the next paragraph we show how this technique applies to the detection of digital lines. We discuss the choice of the set $\{(p, q)\}$ associated with each pixel (i, j) and we present the clustering algorithm.

3.1 Transform from image space to parameter space

A digital straight line is the output of a digitization process applied to a continuous straight line. Digital straight lines have different properties from continuous straight lines. We discuss some of the most important ones below.

1. The slope of an arbitrarily small segment of a continuous line is equal to the slope of the continuous line. This is not the case for a digital line. For example, the slope between two successive points on a digital straight line is either $0, \pi/4$ or $\pi/2$.

2. The structure of the pattern of the pixel arrangement of a digital straight line depends only on the slope of the line.

Property 1 suggests that if two points on a digital line are chosen to represent this line, they should be at a certain distance from one another to give a good estimate of the slope of the line (the value of this distance is a function of the slope of the line). In addition the position of these points in the pattern of the pixel arrangement of the digital line is important.

We take into consideration these two observations in our choice of the set of neighbors (p, q) associated with each processed pixel (i, j), by including in the neighbors set only the points inside a square hollow frame at a given distance centered at the location of the currently processed pixel (see Figure 1). Then we compute and store the parameters (ρ, θ) of the line determined by $((i, j), (p, q))$, for all (p, q). We repeat the whole process for all pixels (i, j). Due to the construction of the algorithm there is only a finite, fairly small (depending on the size of the frame) number of possible angles θ computed in this first step of the algorithm. Amongst all pairs (ρ, θ) we only retain the ones for which the number of occurrences of the same value of θ are high. All other values of θ correspond to points in the image that should be deleted, since they are either isolated or not part of a significant digital straight line.

3.2 Cluster analysis

Among the points (ρ, θ) in parameter space mapped from $((i,j),(p,q))$ in image space there is a spread in the ρ-direction due to the shift introduced by the digital character of the line (see Figure 2). Every digital straight line segment in image space is mapped into a cluster of subclusters in parameter space. Each subcluster consists of points with the same value of θ and closely spaced values of ρ. The median value of ρ in each subcluster varies monotonically as θ increases. The set of values of θ is confined to a finite interval, but the range of values of ρ may become indefinitely large as the length of the digital line segment increases. Thus, in order to detect the presence of a digital straight line segment we need to find clusters in the ρ-direction of a single θ. The technique we use is a one-dimensional single-link hierarchical clustering algorithm [8]. This algorithm uses graph theory techniques to detect clusters. For one given angle θ, we create a symmetric $k \times k$ proximity matrix $D = [d(l,c)]$, where k is the number of different ρ values that are associated with the processed angle θ, and $d(l,c)$ is a measure of the distance between l and c. We then create a threshold graph which is an undirected, unweighted graph without self-loops or multiple edges. Each node in the graph represents a value of ρ. A threshold graph $G(t)$ is defined for each threshold value t by inserting an edge (l,c) between the nodes l and c if the distance between l and c is less than t. That is,

$$(l,c) \in G(t) \quad \textit{if and only if} \quad d(l,c) \leq t \tag{3}$$

The algorithm begins with the disjoint clustering implied by the threshold graph $G(0)$, which contains no edges and which places every value of ρ in a unique cluster. At each step of the algorithm we compute $t \leftarrow t+1$ and we form the threshold graph $G(t)$. If the number of components (maximally connected subgraphs) in $G(t)$ is less than the number of clusters in the current clustering then we redefine the current clustering by naming each component of $G(t)$ as a cluster. The algorithm stops when either $G(t)$ consists of a single connected graph or when $G(t)$ and $G(t+1)$ have the same number of components (it is the user's choice). Once one or several clusters have been detected, we retain the information of minimum and maximum values of ρ for each cluster.

3.3 Refined transform from image space to parameter space

The output of the clustering algorithm defines a set of linear strips in each of which the probability of a line being present is high. Let θ denote the orientation of the medial axis of the strip. Let ρ_{min} and ρ_{max} denote the value of ρ for the two boundary of the strip, $\rho_{max} > \rho_{min}$. Thus the strip in image space is denoted by $(\theta, \rho_{min}, \rho_{max})$. For each of these strips, we associated to every pixel (i,j) inside the strip a set of pixels $\{(p,q)\}$ which are taken only inside the strip and are located further from (i,j) than the original frame width (see Figure 3). All pairs of pixels $((i,j),(p,q))$ are then transformed into a point (ρ, θ) which defines the line passing through the pixels (i,j) and (p,q). We then detect the local maxima of occurrences of the points (ρ, θ), which indicate the presence of sets of collinear points along the line corresponding to the maxima.

3.4 Algorithm

Let $I(x,y)$ denote the digital image. The algorithm implementing this technique is given by :

Step 1 : For each 1-pixel (i,j)

- build a square ring (hollow square region) of length and width $2k$ (where k is a user-set parameter) and of thickness $2 \triangle x$, centered at the location (i,j) (see Figure 11a). We assume that $x_{i+1} - x_i = \triangle x$ and that $\triangle y = \triangle x$.

- for each 1-pixel (p,q) in this region, compute the pair (ρ,θ) defined by the line passing through (i,j) and (p,q),
- store all the pairs (ρ,θ) computed by the above procedure.

Step 2 : Compute the histogram corresponding to the projection of the Hough space onto the θ-axis. By thresholding this histogram, eliminate all the θ values that are not accepted candidate to represent a line.

Step 3 : For all remaining values of θ, labelled θ^r

- apply a 1-dimensional agglomerative algorithm for single-link clustering to the ρ components of the given θ. The proximity matrix is computed using the following distance $d(\rho_i,\rho_j) = |\rho_i - \rho_j|$;
- compute the minimum and maximum values of ρ in each cluster : ρ^r_{min} and ρ^r_{max}

Step 4 : For each cluster

- consider the strip in the image defined by $(\rho^r_{min}, \rho^r_{max}, \theta^r)$, for each 1-pixel (i,j) in this strip build a parallelogram of size $(\rho^r_{max} - \rho^r_{min}) \times w$ at distance $(k + \triangle k)$ from (i,j)
- for each 1-pixel (p,q) in this strip, compute the pair (ρ,θ) defined by the line passing through (i,j) and (p,q),
- store all the pairs (ρ,θ) computed by the above procedure.

Step 5 :
Determine the maxima (ρ,θ) in the parameter space, these values represent sets of collinear points present in the image $I(x,y)$

3.5 Simulation results

Figure 4a shows the original image in which four digital straight lines of different slopes are represented. These straight lines are defined by : $\rho_1 = 47.0635, \theta_1 = 13.4957, \rho_2 = 46.8165, \theta_2 = 69.4440, \rho_3 = 16.96, \theta_3 = -57.9946, \rho_4 = 89.4427, \theta_4 = -63.4349$. Two of the lines have gaps. We apply the algorithm presented in paragraph 3.4 to this image ($k = 7$). The histogram of step 2 is presented in Figure 4b. The local peak values are found to be at $\theta_1 = 15.9454, \theta_2 = 69.4440, \theta_3 = -57.9946, \theta_4 = -66.0375$. We notice that there are also local maxima around the angle values $0, \pi/2$, because they are natural angles for this technique. However, because no line corresponding to either one of these angles is present in the image, the clustering algorithm will eliminate them from the list of possible angles. In figure 4c the result of step 5 is shown in the same histogram form aqs in figure 4b. The location of the peaks are found to be at $\theta_1 = 12.5882, \theta_2 = 69.4440, \theta_3 = -57.9946, \theta_4 = -63.4349$. The error in the value of θ_1 is due to the fact that we used the same $k + \triangle k$ for all the processed angles. Choosing $k + \triangle k$ as a function of θ could resolve this type of errors.

For the ρ-value there can be more than one maximum value due to the pattern created by the digitization process of the line. There are two possible interpretations: the line has a certain width corresponding to the difference of the two extreme values of the maxima, or if we assume that the line is of infinitesimal width, there is a relationship between the values of ρ, the slope of the line known by θ, and the resolution of the image grid. In our case, the lines are considered to be of width 1 pixel; the resulting values of ρ are $\rho_1 = 47.7795, \rho_2 = 46.7590, \rho_3 = 16.8072, \rho_4 = 89.4508$.

4 Conclusion

Preliminary tests indicate that our technique for analysing clusters of points in parameter space while accounting for neighborhood relations in image space, detects and locates digital straight line segments in the presence of noise in image space with good accuracy. One should mention that a single digital line could correspond to any of an infinite number of continuous lines. Thus the highest accuracy that can be achieved remains a function of the digitization process. In the traditional Hough transform, the accuracy of the result is a function of both the digitization of the image space and the quantification of the parameter space. In the method proposed here, the accuracy is only dependent on the digitization of the image space. Our initial tests show that the accuracy of locating straight lines segments by our method is better than that provided by the Hough transform.

Instead of each 1-pixel generating a large number of points in the Hough space (corresponding to all the possible lines passing through the pixel), only a small subset of those points is created in our technique. These points retain the image information of the Hough transform, while reducing by a large factor the data to be processed.

In the traditional Hough technique, if many pixels are concentrated in a small region and are not structured in a line pattern, and if these pixels are all transformed into the same point (ρ, θ), the set of these pixels will be interpreted incorrectly as a digital straight line. Because our technique accounts for local information around each pixel, it will not detect a line in this case. In fact in our technique pixels concentrated in a small region generate distinct values of (ρ, θ), spreading the information all over the parameter space instead of concentrating it at one single point (ρ, θ), thereby reducing the chances of false detection.

References

[1] R.O. Duda, P.E. Hart, "Use of the Hough transformation to detect lines and curves in pictures," Comm. ACM, vol. 15, No. 1, pp 11-15, 1972.

[2] P.V.C. Hough, "Method and means for recognizing complex patterns," U.S. Patent 3,069,654, Dec 18, 1962.

[3] J. Sklansky, "On the Hough Technique for Curve Detection," IEEE Trans. on Computers, Vol C-27, No.10, pp 923-926, 1978.

[4] J. Illingworth, J. Kittler, "A Survey on the Hough Transform," Computer Vision, Graphics, and Image Processing **44**, pp 87-116, 1988.

[5] H. Maitre, "Contribution to the Prediction of Performances of the Hough Transform," IEEE Trans. on Pattern Analysis and Machine Intelligence, Vol. PAMI-8, No. 5, 1986.

[6] T.M. Van Veen, F.C. Groen, "Discretization Errors in the Hough Transform," Pattern Recognition, Vol 14, pp 137-145, 1981.

[7] W. Niblack, D. Petkovic, "On improving the accuracy of the Hough transform," IBM Almaden Research Center Report, No. RJ 6577 (63537), submitted to Machine Vision and Applications, 1989.

[8] A.K. Jain, R.C. Dubes, Algorithms for Clustering Data, Prentice-Hall, pp 60-63, 1988.

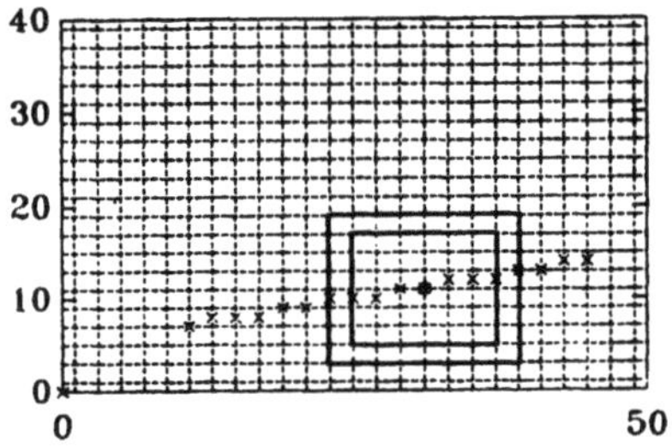

Figure 1: Frame of length and width 6 pixels

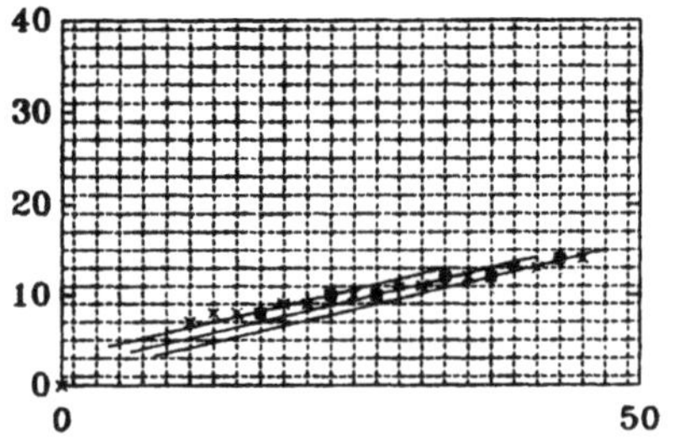

Figure 2: Pictorial representation of the spreading in ρ

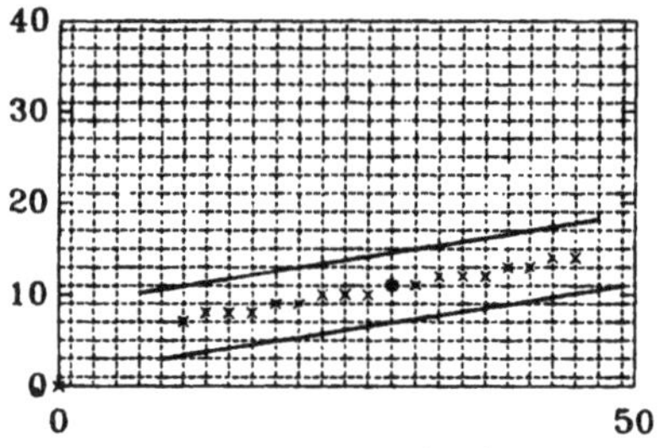

Figure 3: Strip defined by $(\theta, \rho_{min}, \rho_{max})$

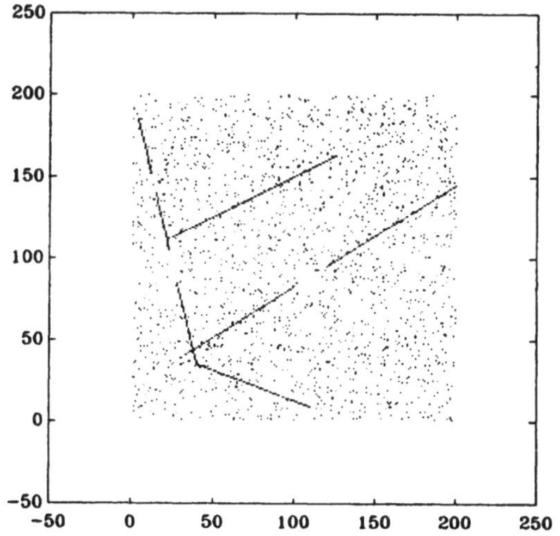

Figure 4a: Original image, Signal:460dots, Noise:2000dots

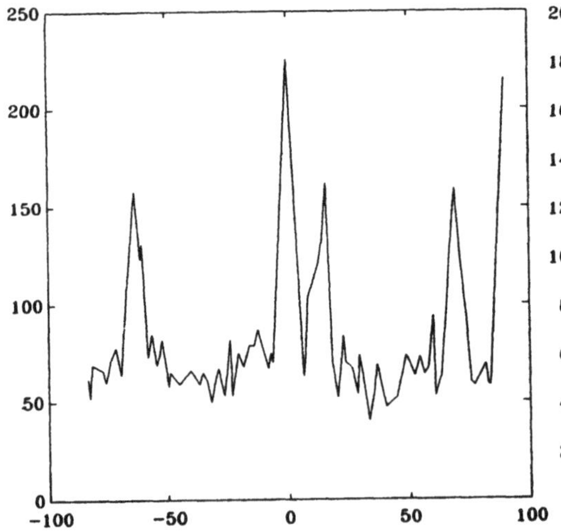

Figure 4b: Histogram in step 2

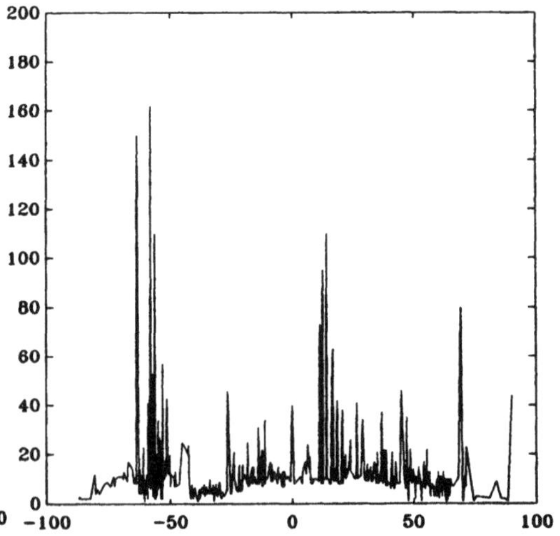

Figure 4c: Histogram in step 5

Figure 4: Illustration of our technique

APPROXIMATION ANALYSIS OF SPACIAL SCENES WITH 3D COORDINATES DATA, OBTAINED BY STEREOTELEVISION PROJECTIONS IN REAL TIME

Syrakov N., Nedev N.[1)]

The visual information has a predominant role in human activities. Therefore its importance will increase with intelligent robots developement of contemporary and future generations. When instead of two dimentions the video information allows the receiving of spacial three dimentional (3D) data with temporary ratios, according to the dynamic changes in the scene the following main advantages are obtained:

- exact information for the location of objects in real time;
- comparative independence from the recurses and relative positions of the videosensors.

The above advantages can be secured if the analysis is implemented on an initial array of 3D coordinates of a certain ammount of points, laying on the visible surfaces of the scene.

In order to obtain the array for 3D coordinates a stereotelevision system with the functional block diagram from fig. 1.b is developped. The video signals U_{VR} and U_{VL} are obtained from the light sensible targets M_L and M_R of a normal orthogonal stereo system with axises orientation explained on fig. 1.a. Solid state CCD video sensors with matrix structure are preffered. The scanning counters S_x and S_z (for axises x and z), constituing the digital equivalents D_x and D_z of the apperture projection coordinates are crystally synchronized with frequency, corresponding of one discrete part scanning of x and generated by the block, marked with the symbol of ⏚. In the blocks S_x and S_z the timing sequencies $\mathcal{C}_x$ and $\mathcal{C}_z$ are constituated also.

As the left (x_L, z_L) and the right (x_R, z_R) projections of an arbitrary spacial point are laying in common epipolar plane, for the adopted at fig. 1.a orientation it is valid:

$$z_L = z_R \tag{1}$$

With every passed frame for $z = z_{max}$ to a delay τ in the scanning of x_R a discrete part of additional delay is added from the block $\Delta\tau$. The latter is opperated from the counter Y. According to the photogrammetry's dependances:

$$[X\ Y\ Z]^t = (b/p)\times[x_L,\ f,\ z_L]\ , \tag{2}$$

where p is the paralax:

$$p = x_L - x_R, \tag{3}$$

1) IMBM, Bulg. Ac. of Scs. Acad. G. Bontchev Bl. No 4, 1113 SOFIA, BULGARIA
Institute of Mechanics and Biomechanics

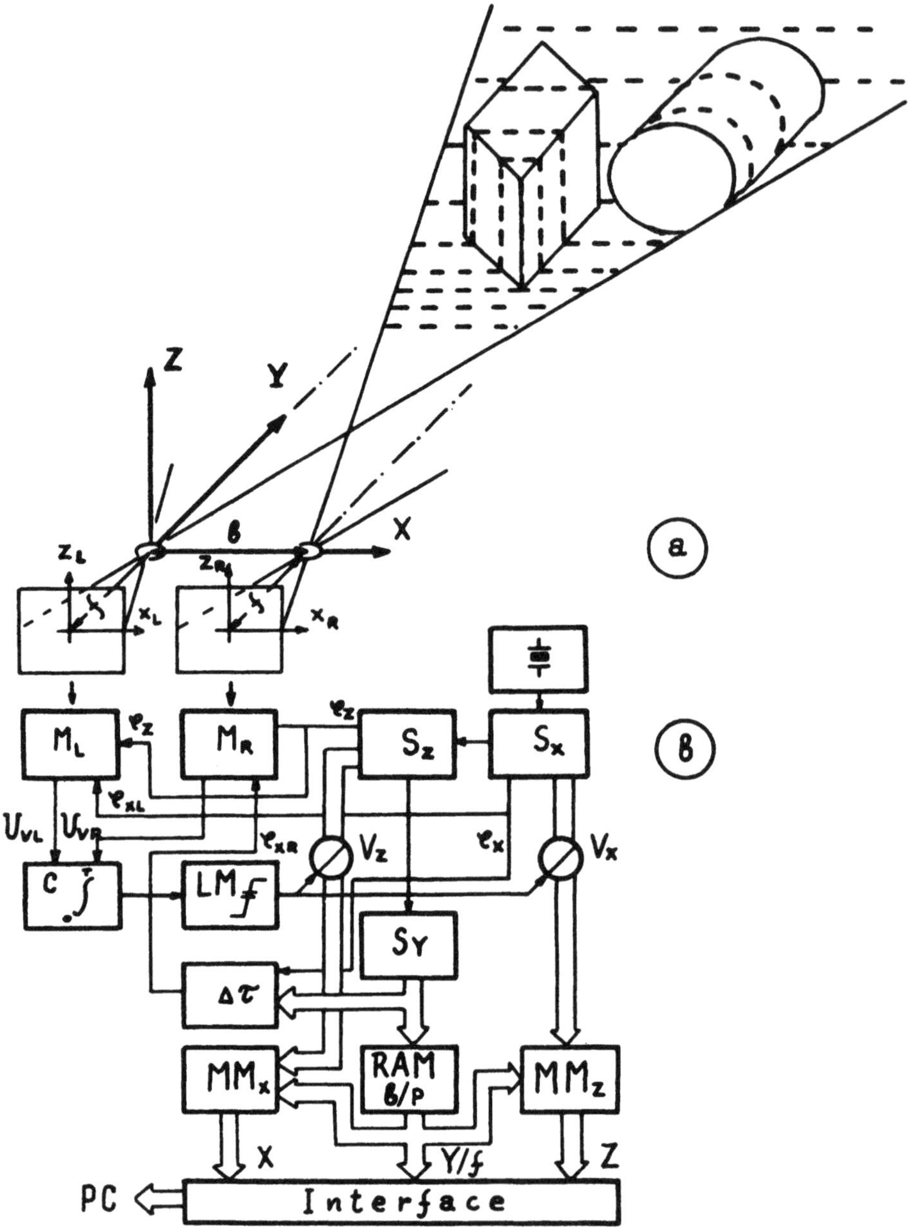
Z
Y
X
b
a
Z_L
X_L
Z_R
X_R
M_L
M_R
S_Z
S_X
b
φ_Z
φ_{XL}
U_{VL}
U_{VR}
φ_{XR}
V_Z
φ_X
V_X
C
LM
S_Y
$\Delta\tau$
MM_X
RAM
b/P
MM_Z
X
Y/f
Z
PC
Interface

Fig. 1.

as $$Y = bf/p \tag{4}$$

for every τ we obtain a scanning of a definite plane Y = const, called in the stereotelevision flat sector. Due to the common epipolar planes (1) in [1] is proved, that stereoidentical points is sufficiant to be searched in identical scanning lines for the adopted orientation. That is why the correlation integral:

$$C_0 = (1/T) \int_0^T U_{VL}(t)U_{VR}(t-\tau)dt \tag{5}$$

will have local maximums of stereoidential projections with well expressed and simmilar changes of dU_V/dt. The correlation interval T is equal to the value of 3 ÷ 8 scanning elements on x and the adaptive interval control is an object of other publications.

Points, matching the right and left projections coordinates from the scanned flat sector are selected by the block LM, detecting in C_0 local maximums with values, prevailing previously statistically investigated and so practically established levels. When a colour TV is used, such opperations have to be executed in parallel in all three RGB channels and the final descision may be found by simple majority logic union. This will increase the reliability, but unfortunately contrast colours rarely exist in industry scenes.

For the so discovered coordinate stereopairs through the latches V_x and V_z the counters states in S_x and S_z are transmitted to the matrix multiplicators MM_x and MM_z. As for the delay τ is valid:

$$k\tau = x_L - x_R = p, \tag{6}$$

in the RAM b/p block can be read the binary equivalent of b/p, which multiplicated by x_L and z_L according to (2) will give the spacial coordinates X and Z, while after an additional multiplication with the camera constant f will give the coordinate Y.

The proposed preprossesor automat is controlled by hardware logic and doesn't require any software. So simmultaniously with the adopted velocity of scanning the 3D array XYZ is obtained, as for every TV frame are described the visible points with relatively maximal correlation C_0 in certain flat sector. This information can be transmitted via fast paralel interface to the memory of the controlling computer PC, where the opperations, exposed below are executed.

For memory reduction and use of integer arithmetics the following organization of the coordinates data is proposed.

Let S_y is the maximal distance on Y, at which the system works with the desired precision, S_x the maximal embracement of the cameras parallel along OX at the distance S_y and S_z - the maximal embracement along OZ at the same distance. Then the space embracement of the cameras can be included in a parallelogram with sides S_x, S_y and S_z, which we denote with B.

Let C_x^j, j = 1,...,n are the minimal distinguished distances along OX in every flat sector and let $C_x = (C_x^1, \ldots, C_x^n)$. In analogy:

$$C_y = (C_y^1, \ldots, C_y^n); \; C_z = (C_z^1, \ldots, C_z^n) \tag{7}$$

If $C = \min(C_x, C_y, C_z)$ the volume B can be fragmented into cubic net with side C .

Then the mapping $\varphi: (XYZ) \to (i,j,k)$, where $i = \left\lfloor \frac{X}{C} \right\rfloor$, $j = \left\lfloor \frac{Y}{C} \right\rfloor$, $k = \left\lfloor \frac{Z}{C} \right\rfloor$, sends the point (XYZ) into a point from the cubic net with integer coordinates.

A bit format is used for preserving the coordinates of the cubic net. The adress of the byte A and the number of the bit NB in the byte, where one is written are defined by the formula:

$$\left| \begin{array}{l} NA + \left\lfloor \frac{i}{8} \right\rfloor + \left(\left\lfloor \frac{n_1}{8} \right\rfloor + \delta\right).(k-1) + \left(\left\lfloor \frac{n_1}{8} \right\rfloor + \delta\right).n_3.(j-1) = A \\ \quad res.\left(\frac{i}{8}\right) = NB, \end{array} \right. \tag{8}$$

where NA is the initial adress in the data memory, and $\delta = \{ {0 \atop 1}$.

Let denote: $R_1 = \left(\left\lfloor \frac{n_1}{8} \right\rfloor + \delta\right)$ and $R_2 = \frac{A-NA}{R_1 \times n_3}$, then from A and NA the coordinates i, j, k are evaluated as:

$$j = \lfloor R_2 \rfloor + 1; \; K = \lfloor res.(R_2)/R_1 \rfloor + 1; \; i = res.\{ res.(R_2)/R_1 \}.8 + NB \tag{9}$$

As $n_1 = \left\lceil \frac{S_x}{C} \right\rceil$, $n_2 = \left\lceil \frac{S_y}{C} \right\rceil$, $n_3 = \left\lceil \frac{S_z}{C} \right\rceil$, the numbers R_1 and $R_x \times n_3$ are are already known and are kept in the memory. Then in (8) are performed two multiplications, 3 additions and one division at 8, in (9) - two divisions, one multiplication and one addition.

It is supposed, that the system, opperates in the set of 3D objects $M = (O_1, \ldots, O_n)$, limited at the bottom by a plane.

Let $O_j \in M$. The series of planes $Y = 0,1,\ldots,n$ from a given place "cuts" the surface O_j into serie of flat figures, laying in the coresponding planes. Depending on the orientation of the object in the coordinate system, a different type of plane figures are obtained.

Let $R = \{R_1, \ldots, R_m\}$ is the space of series, describing the objects of M in all positions in OXYZ.

Let O'x'y'z' is the absolute coordinate system, fixed to a definite point from the working scene. With (P_i, U_j) we denote the location P_i and the orientation U_j of the cameras towards O'x'y'z'.

We suppose that the objects of M are not moving towards each other and towards the cameras.

For every object contained completely in the volume B and every pair (P_i, U_j) we can define the mapping:

$$(P_i, U_j): O_1 \to R_s \tag{10}$$

Let T is the set of all possible pairs (P_i, U_j) $T_1 \in T$ the set of all pairs, for which $O_1 \in B$. The set of pairs, for wich $O_1 \to R_s$ we denote with T_1^s.

Definition 1. If for the object O_1 they are k sets $T_1^{S_i}$ such, that $T_1^{S_i}: O_1 \rightarrow R_{S_i}$ $(i = 1,\ldots,k)$ and $\bigcup_{i=1}^{k} T_1^{S_i} = T_1$ we will define, that the object O_1 has an order of recognition k, and the sets $T_1^{S_i}$ will be define as sets of recognition.

Definition 2. If for the objects O_1 and O_k is fulfilled: $T_1^S: O_1 \rightarrow R_s$, $T_k^S: O_k \rightarrow R_s$, we will define that O_1 is equivalent to O_k relative to R_s. We denote equivalence with $\cong$.

Lemma 1. The objects O_1 and O_k are equals $(O_1 = O_k)$ then and only then, when they have an equal order of recognition k and $O_k \cong O_1$ is fulfilled for k different sets of recognition.

Each object $O_i \in M$ has finite order of recognition r_i, so if $r = \max(r_1,\ldots,r_n)$ from Definition 2 follows, that M is fragmented relative to each serie R_i into sets of equivalent objects.

As the set M is known, for the objects could be defined: the finite series R_i, the sets of recognitions, the order of recognitions and the equivalent objects relative to each serie R_i. On the basis of this information and the introduced consception an algorithm for recognition of the objects from the working scene is developed in [4].

The models of the objects are constructed in the computer memory as finite series of monotone flat figures. For each of the sets of recognition of the objects a corresponding serie is build.

Let the plane figures into which the objects are fragmented are $F = \{f^1,\ldots,f^s\}$. Then the series of the plane figures are build in the following way: $a_1,k_1,m_1,\ldots,a_n,k_n,m_n$; where $a_i \in F$, k_i defines how many times a_i occurs in the serie and m_i is one of the symbols $\{\downarrow,\uparrow,\cong\}$. $\uparrow$-means that the figures are monotonously growing with the number in the serie, $\downarrow$ - the monotonously diminishing and with $\cong$ that they are remaining the same.

Definitions 3. The series $a_1,k_1,m_1,\ldots,a_l,k_l,m_l$ and $b_1,k_1,n_1,\ldots,b_t,k_t,n_t$ are defined as equal, if $l = t$, $m_i = n_i$ for $i = 1,\ldots,t$ and the figures a_i and b_i are simmilar in the plane.

So the problem of recognition of the 3d - objects in M is reduced to recognition of the figures of F in the plane. For solving the problem in the plane the conceptions of regularity and consistency are used.

Let D is a set of opperations and S is a set of conditions.

Definition 4. We will define that $S_i \in S$ is regular relative to $D_j \in D$, if each time when D_j is fulfilled S_i is satisfied. The regularity will be denote with (S_i, D_j).

Definition 5. If M^* is a set of points, we will define that M^* is regular relative to (S_i, D_j), if each time when D_j is fulfilled the points of M^* are satisfying the condition S_i.

If in the plane Oxz we choose the opperation of movement of a straight line x = 0,1,...,n we can determine the regularities satisfacted by the figures of F.

Definition 6. Two regularities will be defined as consistent, if the set of points satisfying them belong to one and the same figure.

It follows from here that the rows of consistency of the figures of F can be determined. Relative to the series of consistancy the set F is fragmented to submultitudes of figures.

On the basis of the introduced conseptions an algorithm for determining of the subsets of figures from F_1 approximating the set of space points, obtained from the stereotelevision system is developed in [3]. The best approximation figure is determined by the criterion for the minimum of the sum of the quadratic errors [2].

As for the objects recognition, such regularities algorithms can be used for a complete 2D scene analysis. As a mather of fact, considering the objects recognition and their coordinates disposition in the scene as a basic problem of the space analysis we can solve many further and more complicated artificial intelligence tasks. We hope, that in its first realizations the system can execute in quasistatic (static for certain time period) real time conditions with a language, allowing work both in primary as in higher levels. Of such a kind are the languages from the FORTH familly, whose openness and optimization abilities we consider as best for the solved problems.

REFERENCES:

1. Nedev N., "Stereotelevision Methods and Systems for Determining the Space Coordinates of Point Marked Moveing Objects", Ph. D. Thesis, HMEI "Lenin", Sofia, 1982.
2. Pavlidis T., "Algorithms for Graphics and Image Processing", Computer Science Press, Inc., 1982.
3. Sirakov N., "Application of regularities in pattern classification", Inter. Con. "Mechanics applications in the Field of Robotics and new Materials", Sunny Beach, 22-26.09.1988, to be published.
4. Sirakov N., "Identification of Objects in R^3 Presented as Final Rows of Plane Figures", First International Conference on Robotic Technologies, 19-26.09.1988. H.V. Diuni, to be published.

A VITERBI ALGORITHM FOR REGION SEGMENTATION AND EDGE DETECTION

I.PITAS

University of Thessaloniki
Department of Electrical Engineering
Thessaloniki 54006,GREECE

ABSTRACT

Two dimensional random Markovian fields have been proposed as stochastic models of real world images. They introduce the stochastic dependence of an image pixel on its neighborhood. The knowledge of the degree of this dependence, its incorporation in a Markovian image model and the use of this model in region segmentation or in edge detection algorithms can improve the results of these algorithms.The present paper proposes the use of the Viterbi algorithm for region segmentation and edge detection based on estimation theory and on two-dimensional Markovian processes.

1. MARKOVIAN IMAGE MODELS

Let x_{ij}, $i=1,..,N$, $j=1,..,M$ be a discrete random field, which corresponds to the pixel (i,j) of an NxM image. This random variable may take the values 1,...,K. The random field x_{ij} is called Markovian random field (MRF), if the probability of x_{ij} depends on the variables x_{kl}, which belong to a neighborhood set D of the pixel (i,j) [9]:

$$p(x_{ij}|x_{11},\dots,x_{NM})=p(x_{ij}|x_{ijD})\triangleq p(x_{ij}|x_{kl}\in D) \qquad (1)$$

Such neighborhood sets are shown in Figure 1. A first order MRF is one for which the neighbor set D is equal to $\{x_{i-1,j},x_{i+1,j},x_{i,j-1},x_{i,j+1}\}$, as it is shown in Figure 1. A second order MRF has neighbor set D consisting of the eight pixels surrounding the central pixel (i,j). We define also as past neighborhood set Dp of an image pixel x_{ij} the pixels belonging to D and satisfying $(k,l) < (i,j)$. Thus the past neighbor set Dp of a second order MRF consists of the pixels $x_{i,j-1}$, $x_{i-1,j-1}$, $x_{i-1,j}$, $x_{i-1,j+1}$. A Markovian field is called separable if

$$p(x_{ij}|x_{ijD})=\prod_{x_{kl}\in D} p(x_{ij}|x_{kl}) \qquad (2)$$

A redundant description of a separable Markovian field is the so-called **trellis diagram** [9]. Part of the trellis which corresponds to the pixel (i,j) and its past neighborhood for a first order MRF is shown in Figure 2, for a two state MRF (K=2). The trellis of the whole image constitutes of the connected trellis diagrams of all its lines. The nodes of the trellis denote the states of x_{ij} and the arcs denote transitions from states of the neighborhood of x_{ij} to the states of x_{ij}. The numbers on the arcs denote transition probabilities.
The use of Markovian fields has already been extensively used in image segmentation and edge detection [8-13]. We have cast the image segmentation in the Markovian model formulation in the following way. If a pixel (i,j) must be labeled by the set $L=(\lambda_1,\dots,\lambda_K)$, the labeling confidence $p_{(i,j)}(\lambda_k)$ can be formulated in the following way:

$$p_{(i,j)}(\lambda_k)=p(x_{ij}=k) \qquad (3)$$

Therefore the problem of image labeling can be stated as follows:

Given an observed image $\underline{y}$, estimate the optimal image labeling $\underline{x}$, when $\underline{x}$ is apriori known to be a 2-d Markovian field.

The edge detection can also be cast in the Markovian model formulation. An edge element is described by its direction, which is aligned with the direction of the maximal local gray-level change and its magnitude which describes the severity of this change. We assume that an edge can have only K-1 discrete directions (e.g. along the axes of 0, 45, 90, 135, 180, 225, 270 and 315 degrees). We denote by x_{ij} a discrete random variable which describes the existence of an edge element at pixel (i,j). This random variable may take the values 0,1,...,K-1. The values 1,...,K-1 denote the possible edge orientation. The value 0 denotes the absence of an edge element in the pixel (i,j). Edges are randomly distributed in an image. Therefore x_{ij} form a two-dimensional random field denoted by a vector $\underline{x}$ of dimension NM, if the image has dimensions NxM. A reasonable edge model is the discrete two-dimensional MRF(4), because, in real cases, the existence of an edge in an image pixel depends only on the existence of an edge in its immediate neighborhood. The problem of edge detection is stated as follows:

Given an observed image $\underline{y}$, estimate the edge distribution $\underline{x}$ of the original image, if it is known that the edges of the original image are a Markovian random field.

In both cases the problem is reduced to the estimation of 2-d Markovian processes. White additive observation noise is assumed. We shall use the Bayesian estimation approach, described in the next section.

2. BAYESIAN ESTIMATION OF 2-D SEPARABLE MARKOVIAN PROCESSES

The joint probability function $p(\underline{x},\underline{y})$ must be maximized:

$$p(\underline{x},\underline{y}) = p(\underline{x})p(\underline{y}|\underline{x}) \tag{4}$$

where $p(\underline{y}/\underline{x})$ is the conditional probability density function of observed image intensity vector $\underline{y}$, given the original image vector $\underline{x}$. Since additive white noise is assumed, the following assumption is reasonable:

$$p(\underline{y}|\underline{x}) = \prod_{i=1}^{N} p(\underline{y}_i|\underline{x}_i) = \prod_{i=1}^{N}\prod_{j=1}^{M} p(y_{ij}|x_{ij}) \tag{5}$$

The probability $p(\underline{x})$ takes the following form, if the model (1) is valid:

$$p(\underline{x}) \triangleq p(x_{NM},\ldots,x_{11}) = p(x_{11}) \prod_{i=1}^{N}\prod_{j=1}^{M} p(x_{ij}|x_{ijD_p}) \tag{6}$$

Therefore the probability $p(\underline{y},\underline{x})$ becomes:

$$p(\underline{x},\underline{y}) = p(x_{11})p(y_{11}|x_{11}) \prod_{\substack{i=1 \\ (i,j)\neq(1,1)}}^{N}\prod_{j=1}^{M} p(x_{ij}|x_{ijD_p})p(y_{ij}|x_{ij}) \tag{7}$$

The optimization of (7) can be performed line-by-line. In such a case, if we try to choose the optimal vector $x_i = [x_{i1},\ldots,x_{iM}]^T$ of the line i, all the optimal vectors $\underline{a}_i$

of the previous lines $l<i$ are known. Therefore, the probability $p(x_{ij}/x_{ijD_p})$ becomes:

(8) $$p(x_{ij}|x_{ijD_p}) = p(x_{ij}|x_{i,j-1},\ldots,x_{i,j-k},\underline{a}_{ijp})$$

for an l-th order MRF. The vector $\underline{a}_{ijp}$ consist of the values of all pixels belonging to the Dp neighborhood set of (i,j) and not belonging the line i. By using the line-by-line optimization the problem is equivalent to the estimation of an one dimensional MRF:

(9) $$p(\underline{x}_i,\underline{y}_i) = p(x_{i1}|\underline{a}_{i1p})p(y_{i1}|x_{i1}) \prod_{j=2}^{M} p(x_{ij}|x_{i,j-1},\ldots,x_{i,j-k},\underline{a}_{ijp})$$

whose transition probabilities $p(x_{ij}/x_{i,j-1},\ldots,x_{i,j-k},\underline{a}_{ijp})$ depend on the already made decisions on the previous lines. If an MRF of order 1 (k=1) is assumed, the logarithm of (9) becomes:

(10) $$\ln p(\underline{x}_i,\underline{y}_i)=\ln p(x_{i1}|\underline{a}_{i1p})+\ln p(y_{i1}|x_{i1})+ \sum_{j=2}^{M}\left[\ln p(y_{ij}|x_{ij})+\ln p(x_{ij}|x_{i,j-1},\underline{a}_{ijp})\right]$$

The following length is attached to each state transition in the trellis:

(11) $$\lambda(x_{ij},x_{i,j-1}) = \ln p(y_{ij}|x_{ij}) + \ln p(x_{ij}|x_{i,j-1},\underline{a}_{ijp})$$

If a separable Markovian model (2) is assumed, (11) takes the form:

(12) $$\lambda(x_{ij},x_{i,j-1}) = \ln p(y_{ij}|x_{ij})+\ln p(x_{ij}|x_{i,j-1})+ \sum_{l=-1}^{1} \ln p(x_{ij}|x_{i-1,j-l})$$

The problem of finding an optimal vector $\underline{x}_i$ is equivalent to finding the shortest route in the trellis of each line, which is solved by the Viterbi algorithm. The Viterbi algorithm is essentially a dynamic programming method. In this paper, we taylor this algorithm to the formulation described by (14). We denote by $\underline{x}_{ik}$ the vector $[x_{i1},x_{i2},\ldots,x_{ik}]^T$ which corresponds to the pixel (i,k). Such a vector denotes a specific path in the trellis, which starts at node 1 and stops at one of the K states of the pixel (i,k). A length $\Lambda(\underline{x}_{ik})$ is associated to this vector:

(13) $$\Lambda(\underline{x}_{ik}) = \sum_{j=2}^{k} \lambda(x_{ij},x_{i,j-1})$$

where $\lambda(x_{ij},x_{i,j-1})$ are the lengths of the transitions which correspond to this path. The vector x_{iM} corresponds to a path in the trellis from node 1 to node M. One of the paths starting form node 1 and stoping at a state of pixel k has the minimal length and it is called **survivor**. The corresponding vector $\underline{x}_{ik}$ is denoted by $\underline{x}_{ik}$. The shortest complete path $\underline{x}_{iM}$ must include one survivor which corresponds to one of the K states of the pixel k. Therefore, if we want to find the complete path $\underline{x}_{iM}$, we can find the survivor of each state of pixel k. Based on these survivors, we can find the survivors of each state of pixel k+1 by adding to the

lengths of the survivors $\underline{x}_{ik}$ the lengths of the transitions from the states of pixel k to the states of pixel k+1 and by choosing the path having the minimal length as a survivor. This procedure can be repeated recursively, until we find the shortest complete path $\underline{x}_{iM}$.
Its formal statement is as follows.

Initialization: k=0, $\hat{\underline{x}}_{i1}=0$, $\Lambda(\hat{\underline{x}}_{i1})=0$

Recursion:
Compute

$$\Lambda(x_{i,k+1},x_{ik}) = \Lambda(\hat{\underline{x}}_{ik})+\lambda(x_{i,k+1},x_{ik})$$

for all transitions $(x_{i,k+1},x_{ik})$.
For each state $x_{i,k+1}$ find x_{ik} such that:

$$\Lambda(\hat{\underline{x}}_{i,k+1})=\min_{x_{ik}} \Lambda(x_{i,k+1},x_{ik})$$

find the vector $\underline{x}_{i,k+1}$:

$$\hat{\underline{x}}_{i,k+1} \triangleq [\hat{\underline{x}}_{ik},x_{i,k+1}]^T$$

The vector $\underline{x}_{iM}$ found gives the optimal estimate of the Markovian randoms field on the image line i.

3. SIMULATION EXAMPLES

The knowledge of the probabilities $p(y_{ij}/x_{ij})$ and of the transition probabilities $p(x_{ij}/x_{i,j-1})$, $p(x_i/x_{i-1,j-1})$, $p(x_{ij}/x_{i-1,j+1})$ is required for the application of the Viterbi algorithm to the image segmentation and labeling problem. These probabilities are not generally known and they have to be estimated from the image itself. Let μ_i, i=1,..,K be the histogram class means for each class λ_i, i=1,..,K. In this case the probabilities $p(y_{ij}/x_{ij})$ are chosen in the following way [4]:

$$p(y_{ij}\ x_{ij}=k) = \frac{1/|y_{ij}-\mu_k|}{\sum_{k=1}^{k} 1/|y_{ij}-\mu_k|} \tag{14}$$

The transition probabilities $p(x_{ij}/x_{i+1,j+m})$ is estimated in the following way [4]:

$$p(x_{ij}=k|x_{i+1,j+m}=k')= \frac{\hat{p}(x_{ij}=k,x_{i+1,j+m}=k')}{\hat{p}(x_{i+1,j+m}=k')} \tag{15}$$

The estimate $p(x_{i+j+m}=k')$ is the average of the probability $p(y_{i+1,j+m}/x_{i1,j+m}=k')$ over the entire image:

$$\hat{p}(x_{i+1,j+m}=k') = \frac{1}{NM} \sum_{i=1}^{N} \sum_{j=1}^{M} p(y_{ij}|x_{ij}=k') \tag{16}$$

The joint probability estimate $p(x_{ij}=k,x_{i1,j+m}=k')$ is given by:

$$\hat{p}(x_{ij}=k,x_{i+1,j+m}=k')= \frac{1}{NM} \sum_{i=1}^{N} \sum_{j=1}^{M} p(y_{ij}|x_{ij}=k)p(y_{i+1,j+m}|x_{i+1,j+m}=k') \tag{17}$$

Viterbi algorithm has been applied to the segmentation of image 3a in four regions. The result is very satisfactory and it is shown in Figure 3b.
In edge detection, the probabilities $p(y_{ij}/x_{ij})$ are chosen in the following, rather heuristic, way. K-1 templates capable of finding edge elements along the K-1 directions are applied to the image. An example of such templates can be the Kirsh operator templates [2]. The template k, k=1,..,K-1 produces an image, showing the edges along a specific direction and having image intensity e_{kij} at the pixel (i,j). The probabilities $p(y_{ij}/x_{ij})$ can be chosen in the following way:

$$(18) \quad p(y_{ij}|x_{ij}=k)=\frac{\max\limits_{k=1}^{K} e_{kij}}{\max\limits_{k=1}^{K}\max\limits_{(i,j)} e_{kij}} \; \frac{e_{kij}}{\sum\limits_{k=1}^{K} e_{kij}}, \quad p(y_{ij}|x_{ij}=0)=1-\frac{\max\limits_{k=1}^{K} e_{kij}}{\max\limits_{k=1}^{K}\max\limits_{(i,j)} e_{kij}}$$

The probability $p(y_{ij}/x_{ij}=0)$ is chosen in such a way, so that the total sum of the probabilities $p(y_{ij}/x_{ij})$ is equal to 1.

REFERENCES

[1] D.H. Ballard, C.M. Brown **Computer vision**, Prentice-Hall, 1982.
[2] M.D. Levine **Vision in man and machine**, McGraw-Hill, 1985.
[3] A, Rosenfeld, R.A. Hummel, S.W. Zucker 'Scene labeling by relaxation operators', **IEEE Transactions on Systems, Man and Cybernetics**, vol.SMC-6, No.6, pp. 420-533, June1976.
[4] S.Peleg, A.Rosenfeld 'Determining compatibility coefficients for curve enchancement relaxation processes', **IEEE Transactions on Systems, Man and Cybernetics**, vol.SMC-8, No.7, pp.548-555, July 1978.
[5] M.D.Levine, S.Shaheen 'A modular computer vision system for picture segmentation and interpretation', **IEEE Transactions on Pattern Analysis and Machine Intelligence**, vol.PAMI-3, No.5, pp.540-556, Sept.1981.
[6] S.W. Zucker, R.A.Hummel, A.Rosenfeld 'An application of relaxation labeling to line and curve enchancement', **IEEE Transactions on Computers**, vol.C-26, No.4, pp. 394-403, April 1977.
[7] E.Wong 'Recursive filtering of two-dimensional random fields' **IEEE Transactionson Information Theory**, pp.84-86, Jan. 1975
[8] D.S.Lebedev 'Probabilistic characterization of images in filtration and restoration problems', **Signal Processing: Theories and Applications (M.Kunt, F. de Coulon editors)**, pp.55-64, North-Holland 1980.
[9] H.Derin, H.Elliot 'Modelling and segmentation of noisy and textured images using Gibbs random fields', **IEEE Transactions on Pattern Analysis and Machine Intelligence**, vol.PAMI-9, No.1, pp.39-55, Jan. 1987.
[10] F.Cohen, D.B.Cooper 'Simple parallel hierarchical and relaxation algorithms for segnenting noncausal Markovian random fields', **IEEE Transactions on Pattern Analysis and Machine Intelligence**, vol.PAMI-9, No.2, pp.195-219, March 1987.
[11] L.L.Scharf, H.Elliot 'Aspects of dynamic programming in signal and image processing', **IEEE Transactions on Automatic Control**, vol.AC-26, pp.1018-1029, Oct.1982.
[12] S.Geman, D.Geman 'Stochastic relaxation, Gibbs distribution and Bayesian restoration of images', **IEEE Transactions on Pattern Analysis and Machine Intelligence**, vol.PAMI-6, pp.721-741, Nov.1984.
[13] D.B.Cooper, H.Elliot, F.S.Cohen, P.F.Symosek, 'Stochastic boundary estimation and object recognition', **Computer Graphics and Image Processing**, pp.326-355, Apr.1980.
[14] G.F.Forney 'The Viterbi algorithm', **Proceedings IEEE**, vol.61, pp.268-278, March 1973.
[15] I. Pitas 'Markovian image models for image labeling and edge detection', **Signal Processing**, vol. 15, pp. 365-374, 1988.

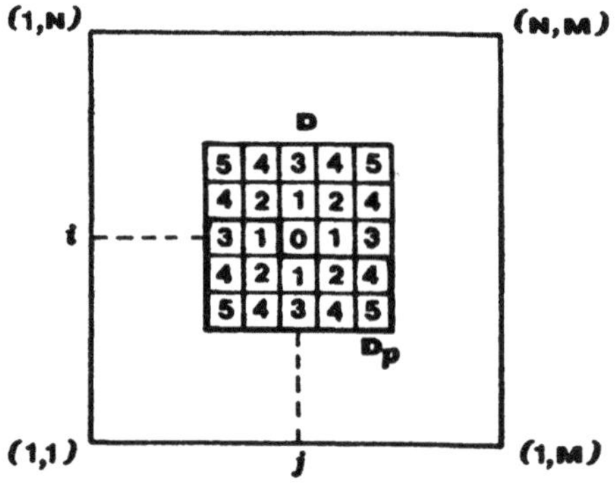

Figure 1: Definition of the Markovian random field. The neighborhood D of the pixel (i,j) consists of all pixels in the square. The past neighborhood D_p consists of all pixels in the boldfaced lower part of the square.

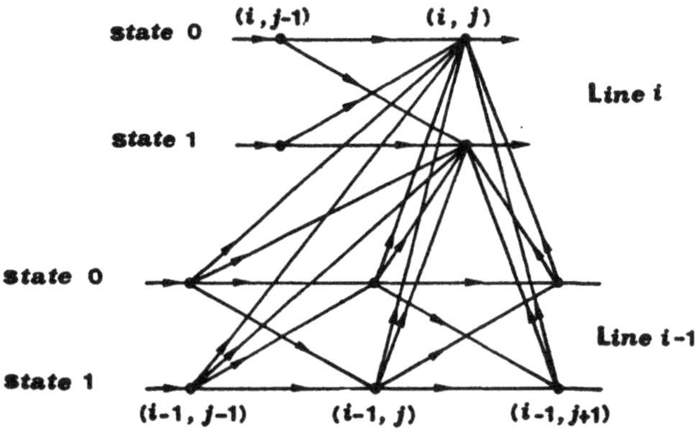

Figure 2: Trellis diagram in the neighborhood of a pixel.

Figure 3: (a) original image
(b) image segmentation in four regions.

A GRADIENT-BASED METHOD FOR GREY-SCALE COMPUTER VISION

V.S.Fain, L.I.Rubanov, M.M.Lutsenko[1] I.Trebaticki[2]

The usual way of formulating of the computer vision (CV) problem consists in restoration of a shape of an arbitrary (opaque) surface from its grey-scale image taken under changeable lighting conditions (LC). The fundamental investigations (see e.g. [1-3]) have shown however that the CV problem in such a statement is not correct and cannot be solved automatically.

In the Institute for Information Transmission Problems of the Sci. Ac. of the USSR (IPPI AN SSSR) in cooperation with the Institute for Technical Cybernetics of the Slovak Sci.Ac. (UTK SAV) the alternative approach to the problem was proposed based on the fact that in a real machine-building industry the number of types of surface shapes is not usually very big [4,5]. Therefore the problem of shape restoration can be substituted by the problem of identification of a type of a surface among the finite set of known types.

The main hypothesis says that the surfaces of the same type must have somewhat similar light distributions. Four most wide-spread types of industrial surfaces were considered: plane, cone, cylinder and sphere. An analysis of many images of real industrial scenes suggested a concretization of the said hypothesis: such characteristics of a gradient of illumination (IG) of an object's surface exist which are determined almost completely by the surface type and are almost independent on LC and on an observation direction.

The following qualitative regularities were revealed:

1) For a plane area of a surface having no patches of reflected light the IG vector has almost homogeneous angle distribution throughout the whole range from 0° to 360° and its module is of rather small average value.

2) For a cylinder area of a surface the IG vector direction is always normal to the cylinder forming line independently on types, locations and other characteristics of light sources (LS) as well as on a quality of surface processing and presence or absence of patches of reflected light. Therefore two directions of an IG vector are possible differing by 180°.

3) For a conic area of a surface the direction of an IG vector is again normal to the forming line of a cone: therefore two possible directions of an IG vector differ by an angle not equal to 180°.

4) For a spheric area of a surface the distribution of an IG vector angle is approximately homogeneous on some finite interval, and the average value of the IG module is rather big.

---

[1] IPPI AN SSSR, Yermolovoa str. 19, Moscow 101447, USSR.

[2] UTK SAV, Dubravska cesta 9, 842 37 Bratislava, Czechoslovakia.

Our purpose in this work is to explain and to ground theoretically those regularities and to confirm experimentally their suitability for a surface shape identification.

The theoretical investigations were conducted for three cases:
(A) IG on object's surface;
(B) IG on an image of an object with diffuse light dissipation;
(C) IG on an image of an object with mirror light reflection.

The LSs were supposed to be remote and point-like for simplicity. It was shown for a combination of several such LSs that the equations describing an IG turn out to be similar to ones connected with a single LS of the type. Therefore the following considerations will be dealing with a single LS without loss of generality.

The set of denotations will be used:
$\mathbf{G}(\mathbf{M})$ - IG at the point $\mathbf{M}$ on an object surface;
$\mathbf{G}(m)$ - IG at the point m on an image of the surface;
$\{\rho,\varphi,z\}$ - coordinates of a point $\mathbf{M}$ in a cylindrical coordinate system;
$\{\mathbf{e}_\rho,\mathbf{e}_\varphi,\mathbf{e}_z\}$ - unit vectors corresponding to a point $\mathbf{M}$ in a cylindrical coordinate system;
$\{\theta,\varphi,\rho\}$ - coordinates of a point $\mathbf{M}$ in a spherical coordinate system;
$\{\mathbf{e}_\theta,\mathbf{e}_\varphi,\mathbf{e}_\rho\}$ - unit vectors corresponding to a point $\mathbf{M}$ in a spherical coordinate system;
$\mathbf{e}_t$ - unit vector normal to a cylinder/cone forming line direction on an image;
$\mathbf{e}_\rho$ - unit vector normal to a line of constant illumination (CIL) on an image;
C_1, C_2, C_3, C_4, C_5 - constants;
b - a parameter of optical projection.

Now consider properties of IG for the cases listed above.

(A) An object surface.

1. For a plane the IG vector is equal to zero everywhere:

$$\mathbf{G}(\mathbf{M}) = \mathbf{0} \qquad (1)$$

2. For a conic surface (in cylinder coordinate):

$$\mathbf{G}(\mathbf{M}) = (C_1/\rho)\cdot f_1(\varphi)\cdot\mathbf{e}_\varphi \qquad (2)$$

where $f_1(\varphi)$ is the function dependent only on φ . It can be seen that the IG vector has constant direction along the forming line (φ = const) of a cone and the module of IG is inversely proportional to the current radius of a cone.

3. A cylinder surface can be considered as a particular case of a conical one if ρ = const. Therefore (see (2)):

$$\mathbf{G}(\mathbf{M}) = C_2\cdot f_1(\varphi)\cdot\mathbf{e}_\varphi \qquad (3)$$

which means that the IG vector has constant module and direction along the forming line of a cylinder (φ = const).

4. The following properties are characteristic for a spherical surface: an absence of lines of constant direction of IG and special shape of any CIL (circle). In the case of a single LS placed in the direction $\theta = 0$

$$\mathbf{G}(\mathbf{M}) = -(C_3/\rho)\cdot\sin\theta\cdot\mathbf{e}_\varphi \qquad (4)$$

The IG module is constant and not equal to zero because the spherical coordinate θ is constant along CIL in this case.

(B) An image of a diffusely dissipating surface.

1. For an image of a plane an IG vector is everywhere equal to zero if a noise can be neglected and random shadows are supposed to be absent:

$$\mathbf{G}(m) = \mathbf{0} \qquad (1')$$

2-3. For an image of a cylindric or conic surface the IG vector can be described by the following expression:

$$\mathbf{G}(m) = C_4\cdot f_2(\varphi)\cdot|\mathbf{G}(\mathbf{M})|\cdot\mathbf{e}_t \qquad (2'\text{-}3')$$

where $\mathbf{G}(\mathbf{M})$ is defined by (2) for conic surface and (3) for cylindric one correspondingly. The direction of IG vector on images of cylinder and cone is normal to the forming line and the module of IG vector is constant along the forming line for a cylindric surface and inversely proportional to a current radius for a conic one.

4. For a sphere image every CIL has special shape (elliptic arc). The expression for the IG vector on a sphere image turns out to be very large, so it is shown below schematically:

$$\mathbf{G}(m) = -C_5\cdot f_3(\theta,b)\cdot|\mathbf{G}(\mathbf{M})|\cdot\mathbf{e}_\rho \qquad (4')$$

where $f_3(\theta,b)$ is a function depending only on θ and b, the last establishing a correspondence between the circle arcs on a sphere and elliptical ones on its image.

It comes from (4') that the IG module is constant along each CIL on a sphere image if observer and LS are at the same location. Otherwise the IG module depends on b.

(C) An image of a mirror-reflecting surface.

This case is naturally somewhat more complicated that the former one because the IG vector of an image depends on an observation direction.

If an observation angle is not equal to the angle between the surface normal in the point and the LS direction then for a surface of any type an illumination of corresponding point of an image from that LS is equal to zero and the IG vector for the surface types considered is described by expressions similar to (1'-4').

Thus we reveal rather simple regularities of light distribution on images of surfaces of the considered set of types. They are resumed in the following table.

IG \ surface	plane	cylinder (forming line)	cone (forming line)	sphere (CIL)
module	0	const	inversely proportional to current radius	const or depends on parameter of optical projection
direction	indefinite	normal to the forming line	normal to the forming line	normal to CIL

Those data confirm theoretically the empirical observations mentioned above (not taking into account a noise component). The regularities revealed gave a base for experiments with real images of scenes of a machine-building kind.

The input of 256×256 pixels' images was made by two manners:
- using the high precision device in the IPPI AN SSSR having mechanical scanning of photo images (256 levels of brightness);
- using the standard TV-camera device in the UTK SAV with a direct input of a scene itself (32 levels of brightness).

The examples of two of the processed images containing cylinders are shown on figs.1,2.

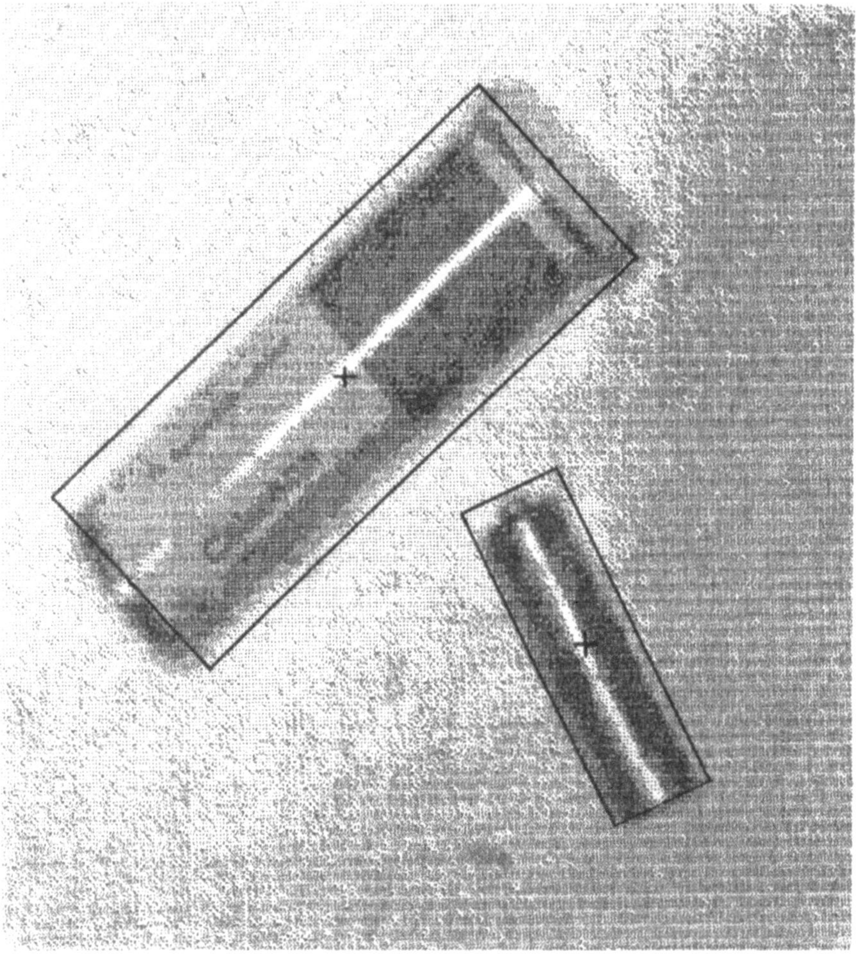

Fig. 1. The first example image.

The digital processing of images was performed by IBM PC/XT computer. The programs were written on Turbo C (version 1.5). There were the following functional groups of program modules:

- preliminary processing (filtering, smoothing etc.) using well-known methods;

- originating (by 3×3 aperture) and averaging (in the window of 8×8) of an IG vector field which leads to the reduced raster of 32×32;

- recognizing of type of a surface shape using the features shown in the table above and evaluating of parameters of a shape (location, orientation, dimensions).

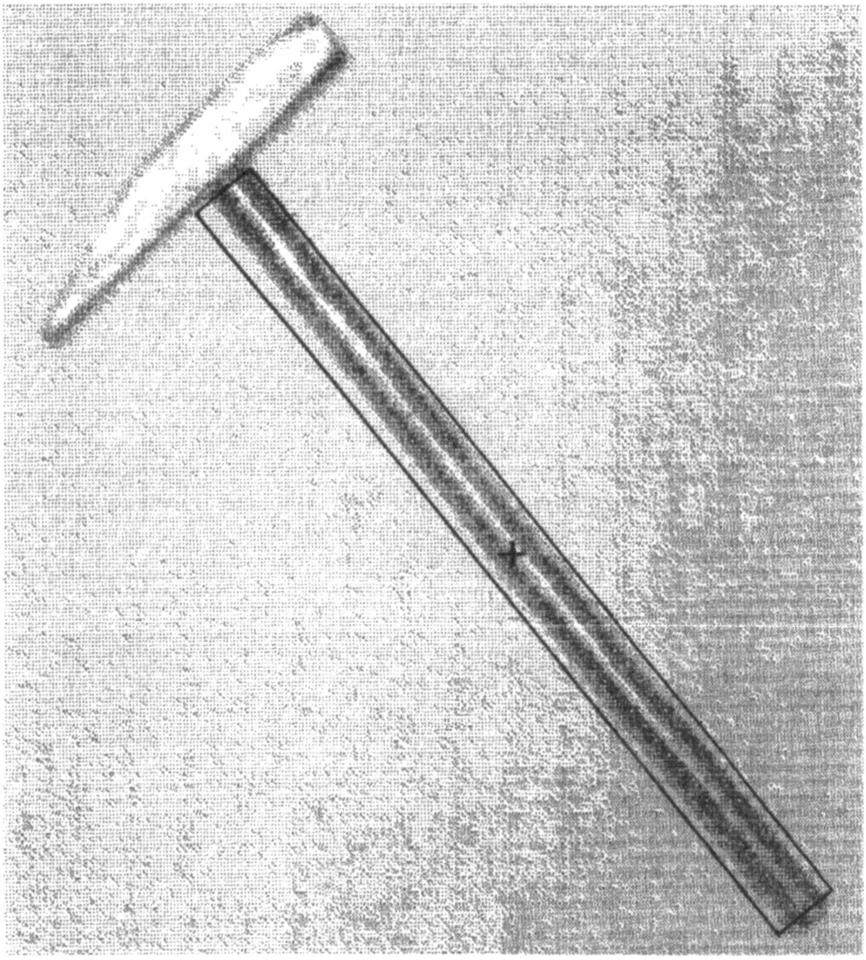

Fig. 2. The second example image.

The data resulting in by these programs' execution are presented on fig.3 and illustrated on figs.1,2 by the solid lines' drawn rectangles. One can see that the coincidence is rather good.

The necessary volume of a memory does not exceed 32 K. The time duration of preliminary processing of an image and reduced gradient field forming takes about 5 minutes (for PC/XT); the time spending for a surface type identification and its parameters determination - no more than 6 seconds. The hardware implementation of the two first groups of program modules is outlined to make the time of performance near to real.

```
Description after angle and direction matching
# 1
Angle of axis = 230
Lenght of cylinder = 170
Diameter of cylinder = 67
X-coord. of center = 104
Y-coord. of center = 104
# 2
Angle of axis = 120
Lenght of cylinder = 97
Diameter of cylinder = 31
X-coord. of center = 176
Y-coord. of center = 172

Number of cylinders = 2
-----------------------------------
Description after angle and direction matching
# 1
Angle of axis = 140
Lenght of cylinder = 261
Diameter of cylinder = 20
X-coord. of center = 144
Y-coord. of center = 152

Number of cylinders = 1
-----------------------------------
```

Fig. 3. Output data of programs' execution.

REFERENCES

[1] Grimson, W.E.L.: An implementation of a computational theory of visual surface interpretation. Comp. Vision, Graph. and Image Processing. 22 (1983) 1, 39-69.

[2] Horn, B.K.P.: Obtaining shape from shading information, In: The Psychology of Machine Vision. McGraw-Hill, NY 1975, 115-155.

[3] Хорн, Б., К. Икеути: Механическое манипулирование произвольно ориентированными объектами. В мире науки. (1984) 10, 76-86.

[4] Файн, В.С.: Распознавание образов и машинное понимание естественного языка. Наука, Москва 1987.

[5] Файн, В.С., Е.К. Вендровский: Существенно полутоновой анализ сцен, В кн.: Кодирование и обработка изображений. Наука, Москва 1988, 144-163.

OBJECT CLASSIFIERS FOR TEXTURED IMAGES

Gintautas Palubinskas [1)]

Abstract. In this work the performance and computer time requirements of 15 classifiers are compared in images modeled by two-dimensional Gaussian Markov random fields which are represented by a causal autoregressive model of the third order. The per-pixel classifier and the object classifiersdirectly or indirectly utilizing spectral-spatial characteristics of image are among them. The probability of misclassification (PMC) calculated analytically and experimentally on modeled data was used as a measure of a classifier performance. The influence of such factors as the object size and form, the inadequacy of a classifier and data models, the accuracy of spatial correlation estimation on the PMC is investigated.

1. Introduction

Consider a two-dimensional multispectral image, where the pixels of an image are q-dimensional vectors, $x = (x^1, \dots, x^q)'$ and x^l is the lth feature of the pixel x. The group of adjacent pixels defines an object of an image. Let us denote the object of an image by X, where $X = (x_1, \dots, x_N)'$ is a vector of size $Nq \times 1$ and N is the number of pixels in the object. Suppose that the pixels of an object have a Gaussian distribution, i.e. $x \sim N(\overset{i}{\mu}, \overset{i}{\Sigma})$, where $\overset{i}{\mu} = (\overset{i}{\mu}{}^1, \dots, \overset{i}{\mu}{}^q)'$ is the mean of the pixel x of the ith class and $\overset{i}{\Sigma}$ is the covariance matrix of x.

The aim of image recognition is to classify each pixel of an object or the whole object of an image into one of m possible classes. In this work decision rules based on the maximum likelihood method are used for solving this problem because they are optimal in the sense of the probability of misclassification.

The classical procedure is a per-pixel classifier which assigns pixel x to the class i, when

$$p(x \mid \omega_i) = \max_{1 \le j \le m} p(x \mid \omega_j), \tag{1}$$

where p is the class-conditional density function for class ω_i. This decision rule classifies pixels alone using only the spectral characteristics of pixels of an image.

Let us have a segmentation of an image into objects. The object classifier [2, 4-8, 10, 11, 13, 14] utilizes the spectral and spatial (class conditional correlation) characteristics of an image. Such a classifier allows decrease of the probability of misclassification and com-

1) Institute of Mathematics and Cybernetics, Academy of Sciences of the Lithuanian SSR, Akademijos str. 4, Vilnius, 232600, USSR

puter time requirements because more information is used to make a decision and fewer décision must be made for the classification of the same image. It must be noticed, that the area of application of the object classifiers is much more wider than that of the classical per-pixel classifier, because additional parameters, such as correlation coefficients between neighboring pixels, are introduced. The decision rule for the object classifier is the following

$$p(X|\omega_i) = \max_{1\le j\le m} p(X|\omega_j), \tag{2}$$

where $X \sim N(\overset{i}{M}, \overset{i}{K})$,

$$p(X|\omega_i) = (2\pi)^{-\frac{1}{2}Nq}(|\overset{i}{K}|)^{-\frac{1}{2}} e^{-\frac{1}{2}(X-\overset{i}{M})'\overset{i}{K}^{-1}(X-\overset{i}{M})},$$

K is the covariance matrix of an object X of size $Nq \times Nq$, $M = (\mu, \dots, \mu)'$ is the mean of X of size $Nq \times 1$.

The usage of the object classifier (2) in general case is very restricted because -for rather large number of pixels in an object N and dimensionality of a pixel q it is difficult to estimate matrix K because of a limited size of the learning sample, and also to calculate $|K|$ (determinant) and K^{-1} even on modern computers. To overcome these difficulties, one has to make certain assumptions about the structure of matrix K. The solution of these problems is based on the two following assumptions about the structure of matrix K.

First, we assume that the correlation between the pixels of an object does not depend on q, i.e. $K = R \otimes \Sigma$, where R is the spatial correlation matrix of size $N \times N$, $\otimes$ is the Kroneker product.

Secondly, we make assumption about the structure of matrix R. Often, it is assumed that the pixels inside an object are independent [5]. In this case R is the identity matrix and the spatial characteristics are employed indirectly. Another popular assumption is that an object of an image is a Markov random field [2, 4, 6-8, 10, 11, 13, 14] which is represented by a causal auto-regressive model. In this case matrix R is characterized by few parameters, the number of which depends on the order of Markov model. The general structure of R is defined by the following formula

$$p(X|\omega_i) = p(x_1|x_2,\dots,x_N,\omega_i)\,p(x_2|x_3,\dots,x_N,\omega_i)\dots \times p(x_{N-1}|x_N,\omega_i)\,p(x_N|\omega_i). \tag{3}$$

For example, for the third order auto-regressive model from (3) we have

$$p(X|\omega_i) = \prod_{j=1}^{N} p(x_j|x_{j1}, x_{j2}, x_{j3}, \omega_i),$$

where x_{j1}, x_{j2} and x_{j3} are right horizontal, right vertical and right down nearest neighbors of x_j, respectively (boundary conditions must be taken into account).

The problem of how to choose the structure of matrix R arises. It is obvious that this problem is data-dependent. In this investigation we have taken data, which occur quite often in the practice, with the following separable correlation function [3]

$$corr(x_{ij}, x_{kl}) = \varrho_1^{|i-k|} \varrho_2^{|j-l|}, \qquad (4)$$

where ϱ_1 and ϱ_2 are spatial correlation coefficients between nearest neighboring pixels of an object in the horizontal and vertical directions, respectively. Images simulated according (4) are called textured images. In this work $\varrho_1 = \varrho_2 = \varrho$.

2. Classifiers

In this work 15 classifiers [11] were designed and then tested on data (4). All of them can be devided into several groups. PIX is the only per-pixel classifier (1) which does not incorporate the spatial characteristics of an image. The other 14 classifiers are object classifiers (the first letter of the name is "O") for an object of size 3 x 3 pixels (except OMARK1 which is for a cross-shaped object of 5 pixels). Five object classifiers incorporate the spatial characteristics of an object indirectly (OPIXVOT, OIND and 3 averaging classifiers: OMEAN, OMEANIND and OMEANMARK3). OPIXVOT is majority vote classifier. Each pixel of an object is classified according PIX and then an object is assigned to the class with the maximum number of pixels of that class. OIND is classifier (2) with independent pixels, i.e. $R = I$. Averaging classifiers classify the mean of an object. OMEAN is minimum distance classifier. OMEANIND and OMEANMARK3 are derived under assumption that pixels independent and the dependence model between pixels is (4), respectively. Nine object classifiers incorporate the spatial characteristics of an object directly (OLINE1, OLINE3, OCROSS, OSTAR1, OSTAR2, OLINE, OMARK2, OMARK3, OMARK1). Classifiers OMARK1, OMARK2, OMARK3 based on causal autoregression of 1st, 2nd and 3rd order, respectively. Other classifiers have intermediate structure of matrix R , from OIND to OMARK3. We have to note, that all object classifiers are easy to generalize for square-shaped and cross-shaped objects (OMARK1) of size $n \times n$, where n is an odd number.

The estimates of the unknown parameters of the classifiers are the maximum likelihood estimates of the classifier based on the causal autoregressive model of 3rd order (OMARK3) [8] . The correlation matrix of this classifier is the same as of the data (4).

3. Classifier selection

In order to select a classifier one has to have a measure of classifier performance (quality). The probability of misclassification (PMC) usualy used as a measure of classifier (1) and (2) performance. There are two ways to evaluate the PMC: experimental and analytical.

In experimental investigation the PMC is evaluated on modeled and real data. But it is very expensive way, because we have to generate data or prepare real data, to write programs for classifiers and then to classify data by these classifiers.

In analytical investigation the PMC is calculated from analytical formulae. In this case we need to have only the mean and the covariance matrices of classifier and data.

3.1. Analytical investigation

For linear classifiers, i.e. $\overset{1}{K} = \overset{2}{K}$, the PMC can be calculated very simply through the distance of Machalanobis.

For quadratic classifiers, i.e. $\overset{1}{K} \neq \overset{2}{K}$, there is no analytical formula for calculating the PMC. It is known several analytical formulae approximating the PMC with the help of distance of divergence and Bhattacharya, Chernoff bounds, etc. But none of these methods let us to compare the PMC of several classifiers on the same data.

In [9] the method of calculation the PMC in the case of classifier covariance matrices and data covariance matrices inadequacy is proposed. It is based on the assumption about the normal distribution of the discriminant function of classifier (2). According [1] the PMC of an object classifier (2) for the two classes case is

$$P = (P_1 + 1 - P_2)/2 , \qquad P_i = \Phi\left[\frac{E(g(X)|\omega_i)}{\sqrt{Var(g(X)|\omega_i)}}\right], \tag{5}$$

where $E(g(X)|\omega_i)$ is the expectation of the discriminant function $g(X) = p(X|\omega_1) - p(X|\omega_2)$ when X is from the class ω_i , $Var(g(X)|\omega_i)$ is the variance of $g(X)$ and Φ is the function of standard Gaussian distribution. Notice, that (5) is the approximation of the PMC for the quadratic classifier. But in [12] it is shown that for $Nq > 30$ ve can approximate the distribution of $g(X)$ by the Gaussian distribution quite exactly.

3.1.1. Linear decision rule ($\overset{1}{K} = \overset{2}{K}$)

In the pioneering work [14] the attempt to investigate the PMC of an object classifier analytically. More recent works are [2, 6] the main result of which shows that the object classifier OMARK3 is much better than the per-pixel classifier PIX.

From (5) we have the PMC of linear classifier

$$P = \Phi\left[-\frac{(\overset{1}{M} - \overset{2}{M})' K^{-1} (\overset{1}{M} - \overset{2}{M})}{2\sqrt{(\overset{1}{M} - \overset{2}{M})' K^{-1} K_T K^{-1} (\overset{1}{M} - \overset{2}{M})}}\right], \tag{6}$$

where $K_T = E((X-EX)(X-EX)')$ is data covariance matrix and K is the covariance matrix of X, on which the object classifier is based.

With the help of the formula (6) the diagrams of the object size dependence on the PMC for four object classifiers OMARK3, OIND, OMARK1 and OPIXVOT for various values of the spatial correlation coefficient ϱ are obtained [11] . Also the diagrams of the PMC of nine object classifiers OIND, OLINE1, OLINE3, OCROSS, OSTAR1, OSTAR2, OLINE, OMARK2 and OMARK3 for various spatial correlation coefficient ϱ values for an object of size 3 x 3 pixels are obtained [11]. Besides that, it is easy to prove that $P_{OIND} = P_{OMEANMARK3} = P_{OMEANIND}$, $P_{OIND} = P_{OMARK1}$ for an object of size 1 x 2 , $P_{OIND} = P_{OMARK3}$ for an object of size 2 x 2 [11].

3.1.2. Quadratic decision rule ($\overset{1}{K} \neq \overset{2}{K}$)

In [2] the Chernoff bounds of the exact PMC of the object classifier OMARK3 are calculated. The main result was that OMARK3 is much better than the per-pixel classifier PIX and the PMC of OMARK3 decreases rapidly with the increase of the size of an object.

Analogically as for the linear decision rule from (5) we have [9]

$$P_i = \Phi\left[\frac{tr\{(\overset{1}{K}{}^{-1} - \overset{2}{K}{}^{-1})\overset{i}{K}_T\} + (-1)^i(\overset{1}{M}-\overset{2}{M})'\overset{\ell}{K}{}^{-1}(\overset{1}{M}-\overset{2}{M}) + \ln(|\overset{1}{K}|/|\overset{2}{K}|)}{\sqrt{2\,tr\{(\overset{1}{K}{}^{-1} - \overset{2}{K}{}^{-1})\overset{i}{K}_T\}^2 + 4(\overset{1}{M}-\overset{2}{M})'\overset{\ell}{K}{}^{-1}\overset{i}{K}_T\overset{\ell}{K}{}^{-1}(\overset{1}{M}-\overset{2}{M})}}\right], \quad (7)$$

where $tr(K)$ is the trace of the matrix K and $\ell = 3-i$, $i=1,2$.

With the help of the formula (7) the diagrams of the PMC of the same nine object classifiers, as in the linear case, for various $\overset{1}{\varrho}$ and $\overset{2}{\varrho}$ values, when $\overset{i}{\varrho}$, $i=1,2$ are estimated exactly, when ϱ of one class is estimated inexactly and when ϱ of both classes are estimated inexactly are obtained [11].

3.2. Experimental investigation

Results of experiments [2, 6] with modeled data illustrate the fact, obtained analytically, that the performance of the object classifier OMARK3 is better than that of the per-pixel classifier PIX.

In [5] the object classifier OIND and the per-pixel classifier PIX are compared on real data. The main result was that OIND is better and faster than PIX. In recent work [4] three object classifiers OMARK3, OIND and OMEAN were investigated on real data. When the classes are not very well separated, OMARK3 has significantly better performance than existing object classifiers. Also, when the classes are moderately well separated, OMEAN which is much faster in calculation, does as well as the others.

In [8] the author compared the performance of three object classifiers OPIXVOT, OIND and OCROSS on real data. The OPIXVOT was the worst and OCROSS was a little better than the other classifiers.

In this work an experimental investigation of the classifiers on modeled data was carried out in order to compare the performance and computer time requirements of all 15 classifiers. For the classifiers OMEAN, OMEANIND, OMEANMARK3, OPIXVOT and OMARK1 for data (4), the analytical PMC was not calculated because of the difficulties in calculating the exact PMC and the PMC of the object classifiers for an object of a different form and also in the situations when the analytical PMC cannot be calculated accurately by the proposed approximate method.

The images were modeled accordingly (4) by an autoregressive equation

$$x_{kl} = \varrho(x_{k,l} + x_{k-1,l}) - \varrho^2 x_{k-1,l-1} + \xi_{kl} ,$$

where ξ_{kl} are independent normal variables.

The image recognition process was carried out in the following way. An image was divided into blocks of size 3 x 3 pixels and these blocks were classified by each of 15 classifiers. The PMC was calculated as the ratio of the number of misclassified blocks to the total number of blocks in the image (error counting method).

The PMC of ten classifiers PIX, OPIXVOT, OMEAN, OMEANIND, OMEANMARK3, OIND, OCROSS, OMARK1, OMARK2 and OMARK3 for various $\overset{1}{\varrho}$ and $\overset{2}{\varrho}$ values are calculated in [11].

4. Conclusions

The performance and computer time requirements of 15 decision making algorithms in images modeled by two-dimensional Gaussian Markov fields were investigated. Random fields were represented by a causal autoregressive model of the third order. The probability of misclassification (PMC) calculated analytically and experimentally on modeled data was used as a measure of classifier performance. The following main conclusions may be drawn.

1. The performance of the object classifiers is much better than that of per-pixel classifiers.
2. The PMC of the object classifiers decreases rapidly with an increase of the size of an object.
3. The performance of linear object classifiers indirectly incorporating spatial characteristics of an object (OCIND) and that of linear object classifiers directly incorporating spatial characteristics(OCDIR) is similar.
4. The performance of quadratic OCDIR is much better than the performance of quadratic OCIND.
5. Among OCIND the classifier OPIXVOT (majority vote) is the worst. The performance of OIND (independent pixels) and averaging classifiers is similar.
6. OCDIR are not very sensitive to the accuracy of spatial corelation estimation and the greater value of the spatial correlation coefficient must be estimated more exactly.

7. The cross-shaped object classifier is better and faster than the square-shaped object classifier for the same number of pixels.
8. Averaging object classifiers are the fastest, followed by OCIND and finally OCDIR.

5. Literature

[1] Fukunaga, K.: Introduction to Statistical Pattern Recognition. Academic Press, New York 1972.

[2] Guyon, X., and J.F.Yao: Analyse discriminante contextuelle.Proc. 5th International Symposium on Data Analysis and Informatics, Paris 1987, vol. 1, 43-52.

[3] Jain, A.K.: Advances in Mathematical models in image processing. Proc. IEEE 69 (1981), 502-528.

[4] Kalayeh, H.M., and D.A. Landgrebe: Stochastic model utilizing spectral and spatial characteristics. IEEE Pattern Analysis and Machine Intelligence 9 (1987), 457-461.

[5] Landgrebe, D.A.: The developgment of spectral-spatial classifier for earth observational data. Pattern Recognition 12 (1980), 165-175.

[6] Mardia, K.V.: Spatial discrimination and classification maps. Communications in Statistics: Theory and Methods 13 (1984), 2181-2197.

[7] Palubinskas, G.: Classification of homogeneous objects in multispectral image data. Statistical Problems of Control. Issue 66, ed. S. Raudys, Institute of Mathematics and Cybernetics of the Academy of Sciences of the Lithuanian SSR, Vilnius 1984, 102-120 (in Russian).

[8] Palubinskas, G.: Spatial image recognition. Statistical Problems of Control. Issue 74, Vilnius 1986, 104-113 (in Russian).

[9] Palubinskas, G.: Probability of misclassification of the Bayes decision rule for normal random vectors in the case of model,data inadequacy. Statistical Problems of Control. Issue 82, Vilnius 1988, 32-42 (in Russian).

[10] Palubinskas, G.; Averaging linear object classifier for image recognition. Statistical Problems of Control. Issue 82, Vilnius 1988, 43-56 (in Russian).

[11] Palubinskas, G.: A comparative study of decision making algorithms in images modeled by Gaussian Markov random fields. International Journal of Pattern Recognition and Artificial Intelligence 2 (1988) 4.

[12] Patnaik, P.B.: The non-central χ^2 and F - distributions and their applications. Biometrica 35 (1948), 202-232.

[13] Sclove, S.L.: Pattern recognition in image processing using interpixel correlation. IEEE Trans. Pattern Analysis and Machine Intelligence 3 (1981), 206-208.

[14] Switzer, P.: Extensions of linear discriminant analysis for statistical classification of remotely sensed satellite imagery. Mathematical Geology 12 (1980), 367-376.

TRADITIONAL TECHNIQUES versus KNOWLEDGE-BASED METHODS in IMAGE ANALYSIS

D.A.Denisov, V.I.Timokhin[1)]

ABSTRACT

The image analysis problem in terms of a problem of visual information transformation from an iconic representation to symbolical one aiming at image descriptions construction is presented in the report. Traditional image segmentation and geometrical descriptions construction techniques are surveyed. The knowledge-based methods for image analysis are opposed to the traditional ones and their advantages are underlined. A new knowledge-based technique for image object identification task is suggested.

1. INTRODUCTION

The image analysis problem in common sense may be treated as a problem of image objects extraction and their descriptions construction. Since starting this kind of research up to the moment much interesting work has been done [1] . The valid basis for the correct comprehension of a problem gives the computer vision methodology [1] . Its fundamental principles are:

1)there exist four levels of videodata representation (iconic images, segmented images, image geometrical structures, image relational structures);

2)there exist correspondence between representation levels, tasks and methods & algorithms (for instance, image restoration and image enhacement tasks are typical for "iconic" level while partition of an image on the homogenous regions - for the "segmented" level and so on;

3)there exists the necessity in knowledge representation and use when solving a particular task;

4)there exists the necessity in use of sophisticated control strategies in order to make videodata transformation process attain the goal.

The formulated "platform" allows to consider image analysis as a compound of image segmentation task and the task of construction of

1)Leningrad Electrical Engineering Institute,
prof.Popov str. 5, 197022, Leningrad, USSR.

objects geometrical descriptions.

In the report the authors try to analyse the traditional methods and algorithms of solving a problem and to point out their restrictions. Our key point is to demonstrate how the knowledge-based approach alters the problem understanding and eleminates most of earlier restrictions.

2. TRADITIONAL IMAGE ANALYSIS

The segmented images are formed from an iconic ones by boundary extraction of homogeneous regions or by labelling their interior pixels. Thus, the main task of segmentation is to find the connected regions of a picture which are uniform with respect to brightness, color or texture. The e d g e d e t e c t i o n and l a b e l l i n g are the dominated processes based on alternative approaches to problem solving. In the first case the solution is provided by pixel properties "discontinuity" if we travel from one region to another; in the second - to pixel properties similarities within the uniform region. The widespread image segmentation techniques such as clustering procedures, relaxation labelling algorithms, region growing methods (including the split and merge algorithms), spartial clustering schemes exstract regions which don't correspond to physical objects. This lacking occurs due to inconsistence between the object model used in algorithm in terms of logical predicate and that complicated that exists in reality. A deep discussion of shortcomings of statistical classification, edge detection and region growing methods was given recently in[5]. The main summary from it may be formulated as following: in order to make the segmentation results more reliable it is necessary to minimize the problem's uncertainty.

The concept-"geometrical structure"-combines the ansemble of the object region geometrical properties. There are two approaches to the geometrical structure description - on a base of features and structural approach. The feature descriptions give us geometrical information by the list of topological, metrical and parametrical features, evaluating connectivity, size, location, orientation, region shape and boundary shape of an object [2]. This list consists of such features as area, perimeter, centroid coordinates, Feret diameters, width, elongateness, thickness of an object etc. Parametrical boundary descriptions may be done by chain code, Freeman contours, Fourier descriptors, spline approximators etc.

The structural descriptions imply disjunction of an object region or its boundary to the primitives and synthesis descriptive structure absorbing various relationships existing between primitives. Therefore, structural approach depicts the visual pattern forming mechanism for a part of an image and an image as a whole as well [3]. Strep trees, concavity trees, connectivity trees, descriptions on a base of medial axes transformation are examples of structural approach utilization [3].

The prominent features of traditional image analysis are:

1)segmentation and geometrical descriptions construction methods are detached from each other;

2)the whole process implies bottom - up control strategy and as a consequence two separate stages of processing: segmentation and description construction;

3)the whole process implies the traditional approach to image object identification by means of the pattern recognition schemes (feature space creation, learning, decision rule formation).

3. KNOWLEDGE-BASED METHODS and TASKS RECONFIGURATION

The knowledge - based methods in general are those, which involve a significant portion of additional information into task solving process. This additional information is saturated with semantics of the domain and is called k n o w l e d g e . The main characteristics of a field is the tasks reconfiguration. In fact, in this case one is not interested in image segmentation and geometrical descriptions construction themselves as it was stated above, but in visual object extraction and their identification.

The knowledge-based methods are minimizing both types of uncertainty: "what to extract?" and "where to extract?" -by utilization objects models during the image analysis process. These models include different kinds of information: grey level peculiarities, shape characteristics, objects spartial relations etc. Thus, two tasks, mentioned above, do not exist appart already, but intertwine in a single framework. Begining from early IGS system and up to the modern SIGMA system these methods find wide utilization in the field of computer vision and image understanding [4]. Nevertheless it can't be said that all problems are taken off from agenda of today's activity.

In this report we are suggesting an image object identification technique on a base of knowledge-based methodology, including relational matching as a decision making mechanism. The main features of a technique are:

1) The set of objects Ω to be identified is arranged hierarchicaly on

a base of "is - a " and "part - whole" relations. Thereby it may be represented by so called "template tree".

2) Object model includes not only the particular feature set and features characteristic values, but computational procedures as well. Every object is characterized by its own feature set and structure. Frame as a knowledge representation formalism is chosen.

3) The matching is controlled by top-down strategy with the "template tree" root and down to the leaves. When matching process arrives to some node of "template tree" the necessary computational procedures are initialised. Then predicate

$$LP\left(\gamma_{\omega_i}, x_k, \Delta_k^{(i)}\right) = \begin{cases} true, & \gamma_{\omega_i} \mathop{\&}\limits_{x_k \in J_i} \left(x_k \in \Delta_k^{(i)}\right) \\ false, & otherwise \end{cases}$$

is computed. Here γ_{ω_i} - boolean feature associated with a tree node ω_i ($\gamma_{\omega_i}=1$ if γ_{ω_j} ; ω_j is a direct predecessor of ω_i); J_i - feature subset for object ω_i ; $\Delta_k^{(i)}$ - characteristic feature values or intervals; x_k - particular features, describing object properties. If $LP(\gamma_{\omega_i}, x_k, \Delta_k^{(i)})$ is true the observed object is identified as it template twin ω_i , otherwise the matching process goes on.

A technique is illustrated by the example of two-dimensional shapes identification. A part of "template tree" is shown on the Fig.1. Some features describing objects properties are presented in the Table 1. Note that the whole number of such variables for approximately 50 objects may exceed 40. The results of application of this techique in some practical tasks would be also reported.

REFERENCES

1. Ballard,D., Brown,C.: Computer Vision. Prentice Hall,N.-Y., 1982.
2. Denisov,D.A., Dudkin,A.K., Pyatkin,V.P.: Digital Image Analysis (Geometrical Structure Descriptions Techniques). Preprint 747, Computer Centre of the Siberian Division of the USSR Academy of Sciencies, Novosibirsk, 1987, (In Russian).
3. Denisov,D.A., Dudkin,A.K., Pyatkin,V.P.: Structural Image Objects Descriptions. Preprint 799, Computer Centre of the USSR Academy of Sciencies, Novosibirsk, 1988, (In Russian).
4. Hwang,V., Davis,L., Matsuyama,T.: Hypothesis Integration in Image Understanding Systems. Comp.Vision, Graph.and Image Proc. 36(1986), 3, 321-371.
5. Wilson,R., Spann,M.: Image Segmentation and Uncertainty. Research Studies Press Ltd., England, 1988.

Table 1

Variable	Semantics	Set of values
x_1	Type of an object parent	The whole set of concepts
x_2	"Contourness" of an object	1, if object is equivalent to its boundary 0, otherwise
x_3	Straightness of boundary	1, the boundary fragment is straight 0, otherwise
x_4	Symmetry of a shape	1, shape is symmetrical 0, otherwise
x_5	Convexity of a shape	1, shape is convex 0, otherwise
x_6	Elongatness of a shape	a, shape is compact b, shape is weakly elongated c, shape is elongated d, shape is strongly elongated
x_7	Concavity measure	a, area of gulfs $\geqslant$ area of object b, area of gulfs $<$ area of object
x_8	Number of acute angles	$\{0,1,2,3,\ldots\}$
x_9	Straightness of medial axes	1, axis is straight 0, otherwise
x_{10}	Type of a gulf of a shape	The whole set of concepts, belonging to type "area"
x_{12}	Existence of "isthmus"	1, yes 0, no
x_{13}	Straightness of medial axes of "isthmus"	1, straight 0, otherwise
x_{17}	Number of sides of a shape	$\{3,4,5,6,\ \ldots\}$
x_{21}	Parallelness of sides	a, two sides are parallel b, four sides are parallel in pairs c, there are no parallel sides
x_{20}	Equality of sides	a, all sides are equal b, parallel sides are equal c, adjacent sides are equal d, absence of equal sides
x_{29}	Existence of self-intersection of a contour	1, yes 0, no

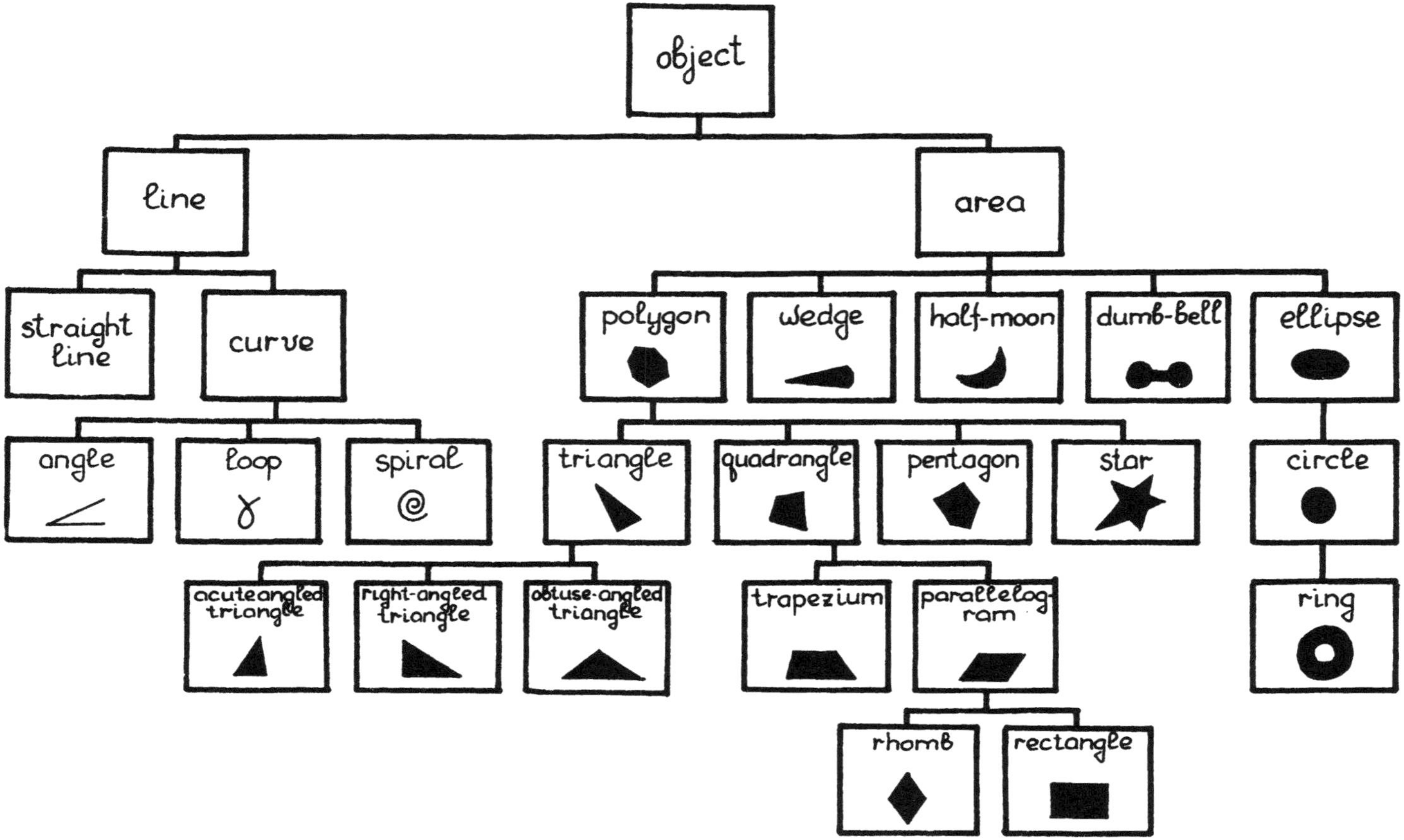

Fig.1 An illustration of a part of hierarchy on a set of shapes.

APPROXIMATION METHODS IN IMAGE SEGMENTATION

René Pischel *)

1. Introduction

Image segmentation is a very important step in image processing, which aim is to separate the image into suitable for further processing regions. Segmentation is an essential prerequisite for all following image processing steps and therefore it is a part of every image processing task. There exists a broad variety of segmentation methods, most of them combine several general signal processing approaches such as digital filtering, data compression, noise reduction and others. Many segmentation approaches, particularly edge-detection techniques, make (implicitly or explicitly) use of numerical differentiation methods. The ill-posed nature of the differentiation problem - even in the continuous case is well-known and leads to a deterioration of the results by noise. Moreover we deal with a discrete function, this results in additional difficulties. One possible way to overcome these problems is to approximate the image function (e.g. by a polynomial) and then to compute the derivatives of the approximating function. In this paper polynomial approximation methods for the 1D - and for the 2D - cases are proposed. It will be shown, that most of the well-known gradient methods implicitly use polynomial approximation approaches.

2. One dimensional approximation methods

The row/columnwise image function $f(i)$, $i=1...N$, defined on the interval $[i_o-m, ..., i_o+m] \subset [1...N]$, can be approximated by polynomials. For fixed i_o (which is omitted in the following formulas for the sake of simplicity, i.e. $f(i_o) \equiv y(i_o) = f_o \equiv y_o$) a general solution for the piecewise approximation using the mean square error method can be described as follows:

*) *Institute for Space Research , Academy of Science, GDR*
Rudower Chaussee 5
Berlin 1199

$$J_m = \sum_{j=-m}^{m} \left[y_j - P(j) \right]^2 \longrightarrow \min \qquad (1)$$

where $P(x)$ is a polynomial of degree p, defined on $[i_o - m, \ldots, i_o + m]$

with $p \leq 2m-1$

Using the state-vector representation

$$\underline{z}(t) = \begin{bmatrix} P(t) \\ \dot{P}(t) \\ \vdots \\ P^{(p)}(t) \end{bmatrix} \quad \text{with } P(t) = \underline{H}_1^T \cdot \underline{z}(t) \text{ , where } \underline{H}_1 = (1, \ldots, 0) \quad \text{and} \qquad (2)$$

$\underline{z}(t) = \Phi(t-s) \cdot \underline{z}(s)$, where $\Phi(t)$ is the transition matrix after differentiation of (1) we obtain a system of linear equations

$$\mathbb{M}_m(k) \cdot \underline{z}(k) = \underline{b}_m(k) \qquad \mathbb{M}_m(k) = \sum_{j=-m}^{m} \Phi^T(j-k) \cdot \underline{H} \cdot \underline{H}^T \cdot \Phi(j-k) \qquad (3)$$

$$\underline{b}_m(k) = \sum_{j=-m}^{m} y_j \cdot \Phi^T(j-k) \cdot \underline{H}$$ and the solution

$$\underline{z}^*(k) = \mathbb{M}^{-1}(k) \cdot \underline{b}(k) \qquad (4)$$

So the approximative first deviation of $y(x)$ yields :

$$y'(i)\Big|_{i=i_o} = y_c \cong P'(0) = \underline{H}_2^T \cdot \underline{z}^*(0) = \underline{H}_2^T \cdot \mathbb{M}_m^{-1}(0) \cdot \underline{b}_m(0) \qquad (5)$$

and particular for linear polynomials $P'(0) = \dfrac{3}{m(m+1)(2m+1)} \sum_{j=-m}^{m} j \cdot y_j$

In our paper various edge detectors which are basing on linear (for m=1,2,3) and cubic (for m=2) polynomial approximation were tested. We also included in our tests an edge detection operator with cubic spline approximation and an operator proposed by H.Jahn [1] ,which for every fixed pixel i compares the left-hand and right-hand limits of the 0. and 1. derivative using left-hand and right-hand linear polynomial approximation. We examined the edge position accuracy, the edge detection accuracy and the error rates of these edge detectors. To derive comparable results we used synthetic data as well as real-world data.

3. Two dimensional approximation methods

Generalizing the 1D approximation algorithms to the 2D case we meet a tremendous manifold of possible extensions of the 1D "sliding window"(i.e. the interval $[i_o - m, \ldots, i_o + m]$). One possibility to reduce this manifold of configurations is to use a region growing scheme. The region growing process starts with an initial region (a set of pixels

(i_k,j_k) k=1...N with the grey values $z_k=f(i_k,j_k)$, which is approximated by a plane $a_N \cdot x + b_N \cdot y + z = c_N$. Now we test all neighboring pixels whether each one of them matches the actual plane in a previously defined sense or not(for instance,if its distance to the plane is too big). The actual plane parameters are uppdated with every new included pixel. The region growing is stopped, if the region is surrounded only by pixels, that mismatch the corresponding plane and then it is started with a new initial region until all pixels are checked. The regions must be connected but they are allowed to have holes. Obviously, the most sensitive and critical point of this method is the criterion, which determines whether a current point $P_n = (i_n,j_n,z_n)$ matches the plane or not, because a best-fit-plane can be formed for any arbitrary set of points. We used here a criterion, which compares the predicted grey value $\hat{z}_n$, obtained from the actual plane $a_N x+b_N y+z = c_N$, with the grey value $z_n = c_{n+1} - a_{n+1} \cdot i_n + b_{n+1} \cdot j_n$, lying in the $a_{N+1}, b_{N+1}, c_{N+1}$ plane, (which includes the point P_n). If the absolute difference doesn't exceed a certain threshold, the current point matches the plane. In some way this criterion turns out to be a generalization of the comparison criterion of left-hand and right-hand limits of the 0. and 1. derivative of the 1D intensity function, proposed by H.Jahn [1].Here the right-hand limit corresponds to the prediction value.

4. Conclusion

Although approximation methods are used for 1D profiles of the intensity function (i.e. row/columnwise) rather than for the 2D case it is promising to connect the advantages of approximation methods such as their nonheuristic nature and their simplicity with the full directional information, which is available from the 2D image function. Independently of that, applying linear approximation methods one has to bear in mind the underlying piecewise linear image model, because differences between the model and the real image function of course would lead to uncertain solutions.

References

[1] *Jahn, H. Edge detection on the basis of piecewise linear models*
Proc CAIP '87 ,Wismar 1987 , p.23
Series "Mathematical Research", vol. 40 , Akademie-Verlag, Berlin

AN ALGORITHM FOR THE RECOGNITION OF LINE OBJECTS

Schirmer, H.; Fuchs, S.[1]

1. Introduction

A large class of pictures to be processed automatically is characterized by a high amount of line structures (e.g. drawings, maps).
The recognition of line objects is a difficult problem with regard to the following aspects:

- there are line objects of different kinds, e.g. small and heavy full lines, broken lines etc.
- line objects touch and intersect one another and other objects.

The most important methods suggested in the recent literature for analysing the structure of images such as the isomorphism of labeled subgraphs or the consistent labeling are not suitable for recognition of line objects. These methods are designed for other problem statements, namely for the recognition of objects composed of various parts (regions) of the image having certain features and relevant adjacency relations. However line structures consist of only one region the adjacency relations of which be unessential for the recognition. But the methods named may be useful for the succeeding semantic process, as e.g. the recognizing of complex objects. To generate features of line structures one has to examine geometrical relations of the image. The image representation in form of vectors supports the computation of geometrical features better than the raster image does. The data structure suggested by Kovalevsky [1] the so-called cell list structure based on the topological concept of cellular complexes is an efficient means for the structural analysis. An example of this is being given at this conference in [2]. The so-called block cell complex consists of cells of different dimension. The 2-cells are connected regions of elements (pixel) having a constant label. The boundary of the 2-cells are the 1-cells which may have 0-cells (branching points) at the locations where three or more regions meet.
The structure of the cell list is the following.
The list of 0-cells contains the coordinates of the branching points and the pointers to all 1-cells (lines) bounded by the 0-cells.
The list of 1-cells contains the pointers to the starting and end point of the line (0 if the line is closed), pointers to the 2-cells (regions) lying to the right and to the left hand side of the line, and pointers to the begin and end of the subsequence of the metric list.
The list of 2-cells contains the gray value of the regions, and the pointer to one of the lines bounding the region.
The list of the metric data of the 1-cells contains the data of polygon representation of the lines. These are coordinates of the end points of the digital straight sections (DSS), and additional parameters for the reconstruction of the original boundaries.

2. Representation of the method and the algorithm

2.1. The suggested concept

The aim is to split the process of recognition into such steps that are independent of the special applications and into other steps using the knowledge about the contents of the image (Fig. 1):
Step 1: Segmentation

---

[1] Informatik-Zentrum des Hochschulwesens an der Technischen Universitaet Dresden
Mommsenstr. 13, Dresden, 8027, DDR

Step 2: Generation of cell list

Step 3: Search of parts of line objects without branching points

Step 4: Connection of the parts of line objects

The result is the polygon of the middle axis of the line object which may contain branching points. This structure corresponds to a 1-dimensional cell complex presented as a sequence of 0- and 1-cells. Additional to the cell list of the original image the 1-cells of line objects contain attributes e.g. the gray value and the breadth of the object.

Step 5: Semantic image analysis

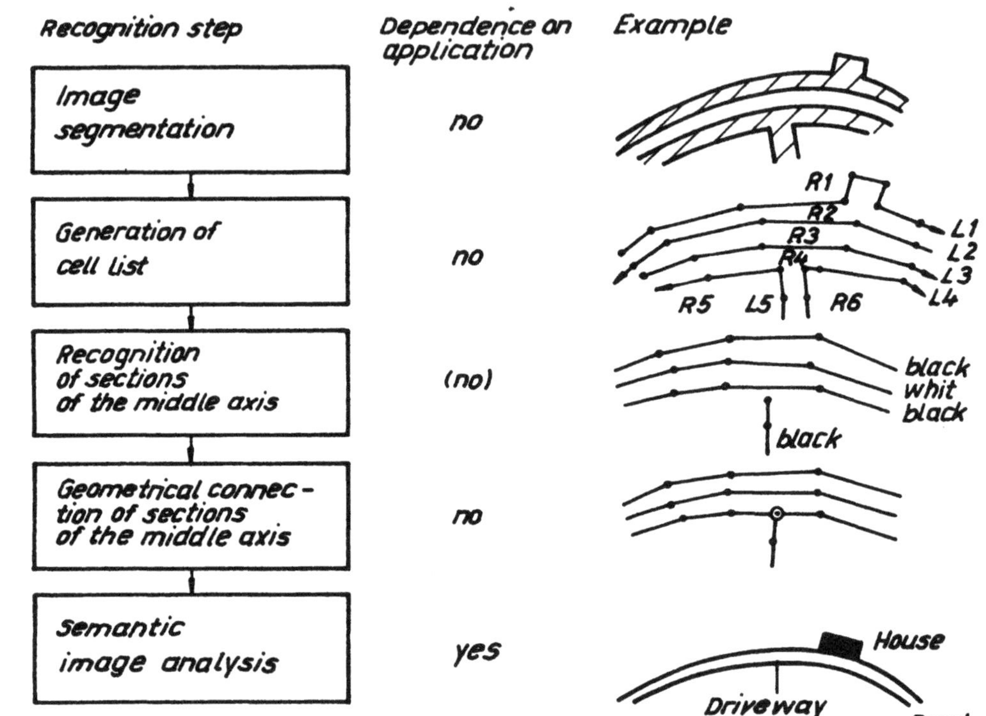

Fig. 1 : Concept of the recognition process

2.2. Techniques for recognizing the parts of line objects (step 3 of the concept)

Method 1: The well-known method of the bisectrix is used to generate the middle axis of the line object (Fig. 2). Starting at any entry point the middle axis is formed by sections of bisectrix of two boundary sections lying oppositely. Each of such straight sections is bounded by the intersection with the bisectrix of two adjacent boundary straights. The advantage of this method is the accuracy. Of course, it is difficult to find all entry points and to continue the line tracing at locations where the line object is disturbed.

Entry points be free ends of lines, ends at the image boundary, ends at other objects, and branching points of line objects.

To continue the line tracing at the "trouble section" we suggest the following algorithm:

- there is a trouble section, if the breadth of line object is too small or too large
- that side of boundary lines is found out where the angle between two adjacent straight segments is large
- on this side a new intermediate point is searched in the old direction with a giver

tolerance of angle.
The algorithm breaks down, if there are both trouble section and changing the direction of the line object.

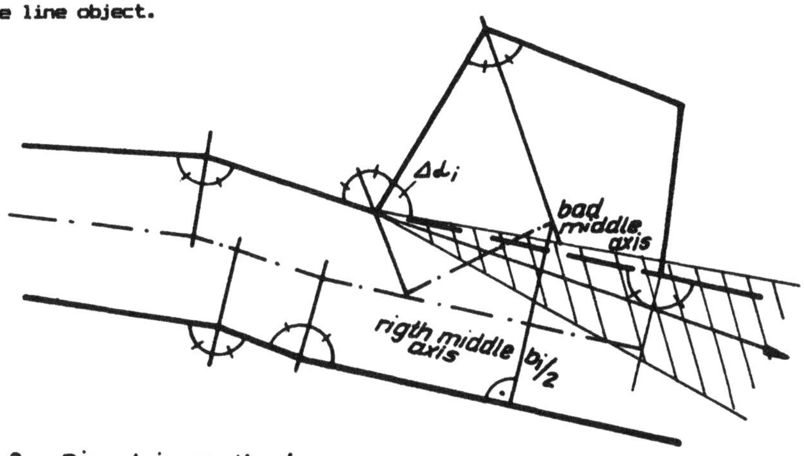

Fig. 2: Bisectrix method

Method 2: To overcome this difficulty we suggest a new method employing a more global strategy. The idea consists in the following.

(1) All the pairs of digital straight segments being potential boundaries of line structures are found out (Fig. 3). (DSS_i, DSS_j) compose a pair, if the elements satisfy the following conditions:

   (a) DSS_i, DSS_j are to be boundary straights of the same region R_K having the given gray value GV. They must be lying oppositely.

   (b) The difference of their angles must be small.

   $|\alpha_i - \alpha_j| < \Delta\alpha$

   (c) There exists at least one perpendicular from one of the four points to the straight lying oppositely. The length of this perpendicular b_{ij} that corresponds to the breadth of the line object must be in the range

   $b - \Delta b \leq b_{ij} \leq b + \Delta b$

   where b is the breadth and Δb is the tolerance of the breadth.

   (d) The lengths of the DSS must not be too small

   $l_i, l_j \geq l_{min}$

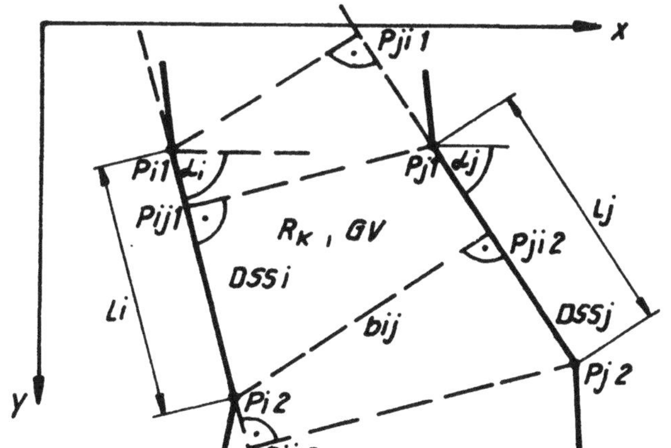

Fig. 3: DSS pairs (DSSi, DSSj)

(2) The current breadth may be calculated as the mean value of the breadths of all the pairs containing DSS_i

$$b_i = \frac{1}{n_i} \sum_{j=1}^{n_i} b_{ij}$$

with n_i = number of pairs with DSS_i.

(3) The boundary lines L_m of the region R_K are weighted and sorted with regard to the length of potential boundary of line objects (Fig. 4). We will sort the boundary lines using the quantity

$$\frac{l_{pm}}{l_{am}} \left(1 + \frac{l_{pm}}{l_p}\right)$$

where l_{pm} is the sum of the lengths of assigned DSS_i of the line L_m,
l_{am} is the sum of the lengths of all DSS_i of the line l_m, and
l_p is the sum of the lengths of all assigned DSS_i of all boundary lines of the region R_K.

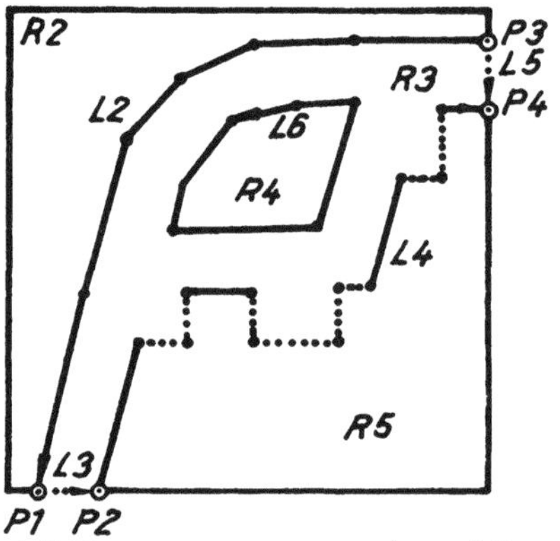

Sequence of the weighted lines of the region R3 : L2, L6, L4, L5, L3

Fig. 4 : Weighting and sorting the boundary lines of a region

(4) The middle axis of the line objects is constructed using the boundary lines in the sorted sequence as basic lines (Fig. 5)

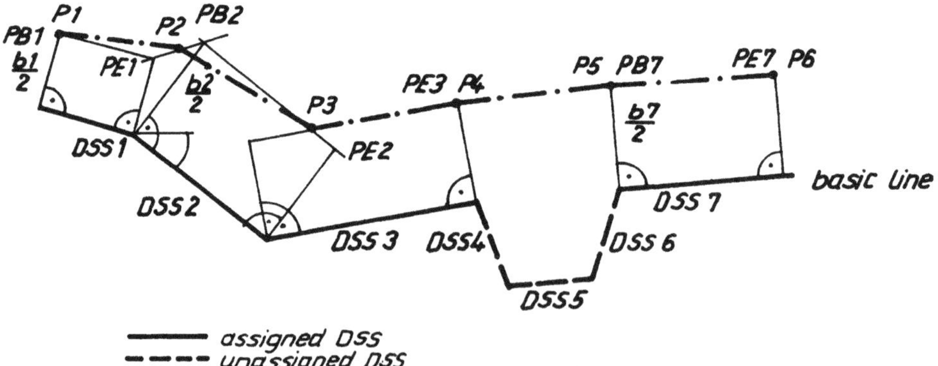

Fig. 5 : Generation of sections of the middle axis (SMA)

Each assigned DSS_i of the basic line generates one straight lying within the region R_k and having the distance $b_i/2$ from the DSS_i. The pairs of DSS already processed are removed from the listing, because they must not generate a middle axis once more, when another basic line is being processed. To provide the polygon representation of the middle axis, it is necessary to update the end points of the straights.

We shall do this in the following way:

(a) DSS_i, DSS_{i+1} be assigned and adjacent

case 1: $|\alpha_i - \alpha_{i+1}| \leq \alpha_0$

The end point P_{Ei} of the straight i and the starting point P_{Bi+1} of the straight i+1 are replaced by their centre P_n.

case 2: $|\alpha_i - \alpha_{i+1}| > \alpha_0$

The end point P_{Ei} and the starting point P_{Bi+1} are replaced by the intersection P_n of the straights

(b) DSS_i, DSS_{i+r} be assigned but not adjacent

The end point P_{Ei} and the starting point P_{Bi+r} are labeled as intermediate points of polygon. The gap will be closed by the connecting straight.

If any straight of polygon of the middle axis is intersecting any boundary line this straight must be removed and the polygon must be interrupted at this location. A new section of the middle axis is opened.

2.3. Geometrically connecting of sections of middle axis (step 4 of the concept)

The goal of this step is to connect various sections of the middle axis of the same region and to recognize the branching points. The connecting line is always beginning at an end point of a section of the middle axis. Any sections of middle axes (SAM) are connected only, if

(1) the smallest distance from the end point of the SMA to another SMA is not greater than a given threshold

(2) the connecting line is not intersecting boundary lines.

As shown in Fig. 6 we distinguish between two kinds of connecting lines:

(1) The nearest point is also an end point of a SMA.

If $|\alpha_i - \alpha_j| \leq \alpha_1$ then the two points are simply connected by a straight else both straights at the end of the two SMA are extended up to their intersection. No branching point is provided.

(2) The nearest point is not an end point of a SMA.

If $|\alpha_i - \alpha_{jn}| \leq \alpha_1$, where n=0,1 or 2, then the end point is connected with the point P_{jn} of SMA_j. The point P_{jn} is set as a branching point.

If $|\alpha_i - \alpha_{jn}| > \alpha_1$, where n=0,1,2, then the end straight of SMA_i is extended up to the intersection P_{Bij} with the SMA_j. The intersection is set as branching point.

3. Test of the suggested algorithm

The algorithm has been implemented on the computer "BVS Robotron A 6472" in the FORTRAN-77 programming language, where the cell list is generated using the special processor DIP. The algorithm has been tested by analysing digitalized printed maps containing different streets overlapped with other objects. The experiments have shown that 95% up to 100% of the length of the line structures were correctly recognized. The computer time for analysing an 128x128-image with the computer Robotron K 1630 without arithmetic processor was about 60 s. On computers with floating point operations the computer time may be substantially reduced.

A problem which is not completely solved up to now is the recognition of star-shaped crossing-points.

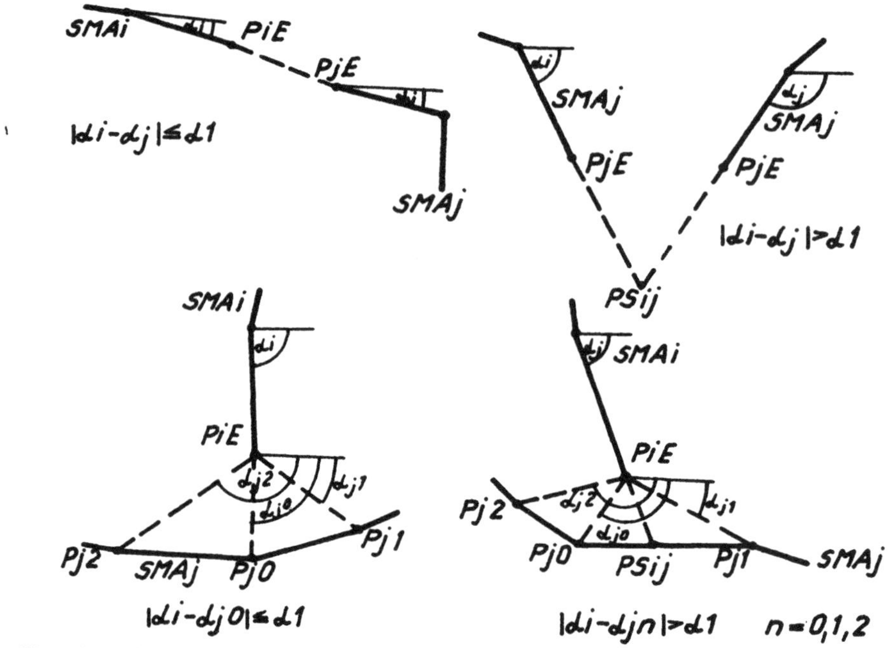

Fig. 6 : Geometrical connetion of sections of middle axes

4. Conclusion

The suggested method is suited for recognizing line structures of different kinds. It provides image parts which may be used to interpret the semantic contents of the whole image. Thus the method may contribute to "computer vision" and "image understanding".

References:

[1] Kovalevsky, V.A.: Finite Topology as applied to image. Analysis, CVGIP 1989 (to be printed)

[2] Fuchs, S.; Kovalevsky, V.A.: Method for the structural recognition of noisy patterns. CAIP '89.

EFFECTIVE LINE-BY-LINE ALGORITHM FOR SKELETONIZATION OF DIGITAL PICTURES

V.M. Kiyko[1)]

Introduction. Let the picture $V \subset R^2$ be the subset of points on an Euclidean plane, every connected component of the contour(boundary) of this picture being closed poligonal line without selfcrossings, composed of the finite number of horizontal and vertical line segments. Certain square $S_M \subset R^2$ is called maximal square if a) sides of this square are the segments of vertical and horizontal lines; b) $S_M \subset V$; c) there is no square $S \neq S_M$ for which conditions a), b) and $S_M \subset S$ are satisfied simultaneosly.

The basic skeleton of the picture V is defined as the set of central points of the maximal squares. Every connected component of the basic skeleton (Fig. 1) is either an isolated point, or consists of one or more horizontal and vertical segments. There are two types of the ends of the basic skeleton segments: the close end if it is the central point of the maximal square and the open end otherwise. Let t_o be the open end of the segment of the basic skeleton. The maximal square S_M is defined as corresponding to point t_o one, if S_M contains all squares $S \subset V$ with the central point t_o . The auxiliary skeleton is the set of additional line segments, each of them linking open end of the basic skeleton segment with the central point of the corresponding maximal square. The basic and auxiliary skeletons are shown in Fig. 1 by thick and thin lines accordingly.

Our goal is to develop the effective algorithm for detecting the basic and auxiliary skeletons in any binary picture V .

Special features of the algorithm. Let vertices of those polygons that constitute the contour of the picture be called corners. All eight types of the corners are shown in Fig. 2, corners of types 1 - 4 being called convex and others concave.

Let q_o be the concave corner of the picture V . Certain square is called dissective square (Fig. 3) if a) $S_p \subset V$; b) one of the vertices of the square S_p coincides with the corner q_o ; c) one of the diagonals of the square S_p coincides with the bisectrix of the corner q_o ; d) there is no square $S \neq S_p$ for which conditions a), b), c) and $S_p \subset S$ are satisfied simultaneously. Due to last feature of the square S_p at least one of its two sides, containing no q_o , has non-empty intersection with the contour of the picture V . Let q^* be any point of this intersection. Let us suppose also that points q_o and q^*

1) Academy of Sciences of the Ukrainian SSR, V.M. Glushkov's Institute of Cybernetics, Acad. Glushkov prosp. 20, Kiev, 252207, USSR.

belong to the same connected component G_c of the contour of the picture V. In this case points q_0 and q^* split the contour G_c into two parts G_{1c}, G_{2c} and boundary of the square S_p into two parts G_{1s}, G_{2c}. Splitting the picture V into two pictures V_1 and V_2 (Fig. 3) is called a decomposition if a) $V_1 \cap V_2 = S_p$ and b) the contour of each picture V_1 or V_2 contains one and only one of the parts G_{1c}, G_{2c} and only one of the parts G_{1s}, G_{2s}. Let us define also a connector as the line segment linking central points of two maximal squares containing the square S_p and belonging to the pictures V_1 and V_2 correspondingly. The following two statements are valid.

<u>Statement 1</u>. The basic skeleton of the picture V is the union of the basic skeletons of the pictures V_1 and V_2.

<u>Statement 2</u>. The auxiliary skeleton of the picture V is the union of the auxiliary skeletons of the pictures V_1, V_2 and a third set containing only the connector.

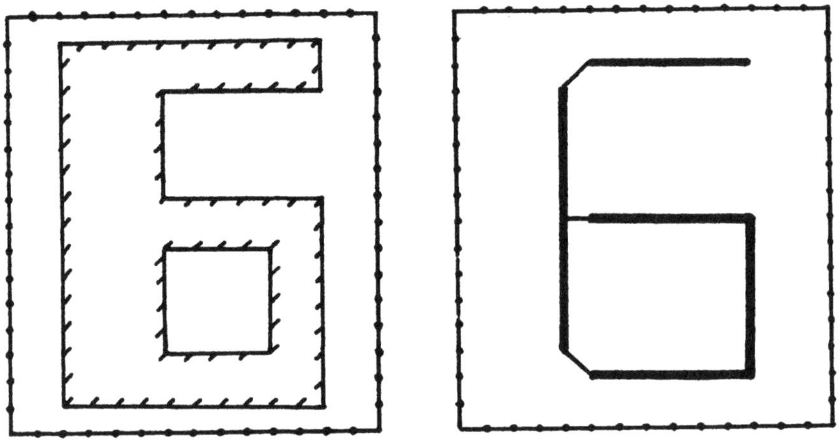

a b

Fig. 1. The basic and auxiliary skeletons (b) of the picture (a).

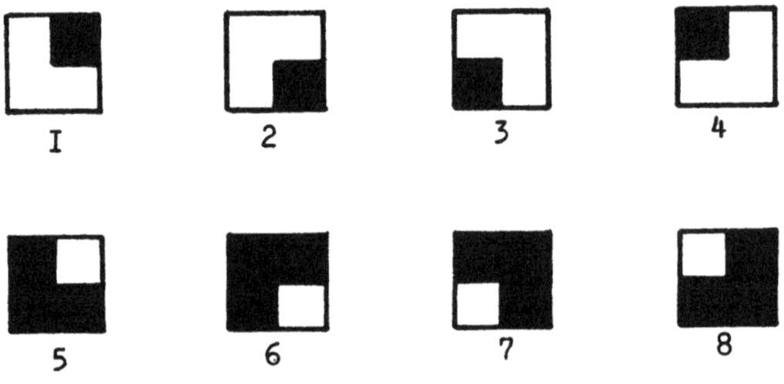

Fig. 2. The types of the picture corners.

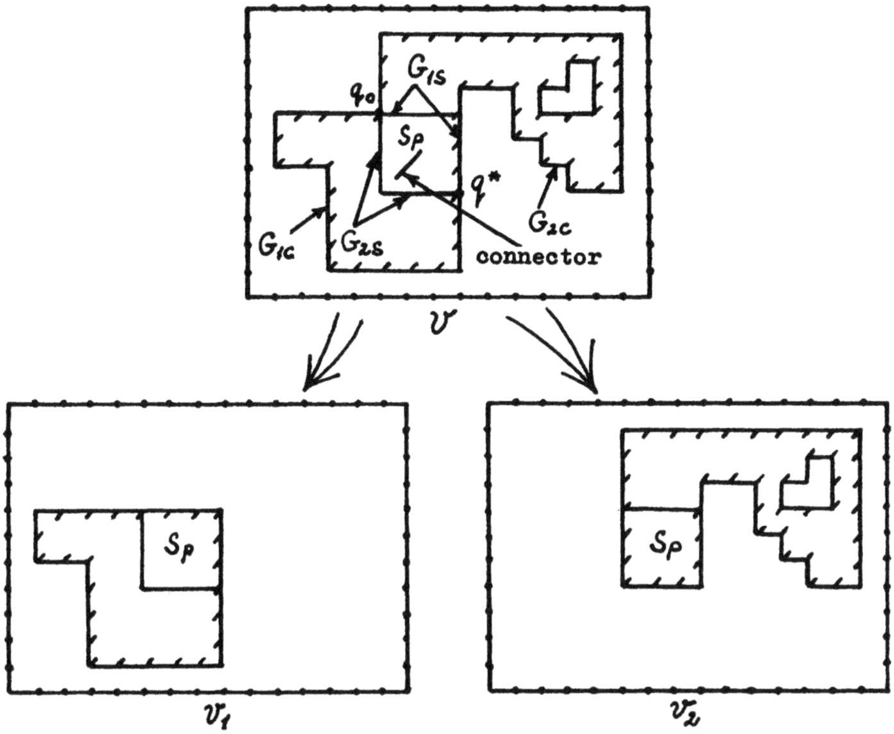

Fig. 3. Decomposition of the picture V into two pictures V_1 and V_2 .

Certain picture is called a simple picture if its contour contains no corners of the types 5 and 8. The contour of every connected component V_K of the simple picture (Fig. 4) contains one corner of the type 2 and one corner of the type 3. If the picture V_K contains no concave corners, then it is a rectangle and its skeletonization is obvious. Otherwise the skeleton of the picture V_K can be detected in the following manner. The concave corner $q_0 \in V_K$ with the least coordinate x and dissective square corresponding to this corner are determined. Then the decomposition of the picture V_K into two parts is carried out, one of these parts being the rectangle and other being the simple picture V_{K1} with no concave corner q_0. These operations are repeated up to complete removal of concave corners from the picture V_{K1}.

The algorithm below consists in succesive splitting of any picture into simple pictures. This algorithm posseses two main features in comparison with the known algorithms [1,2]: its computational complexity depends on no thickness of the picture and is determined by no number of the pixels but the essentially less number of the picture corners. The additional positive feature is that the algorithm requires comparatively small main memory of the computer depending on no length of the scanned part of the picture.

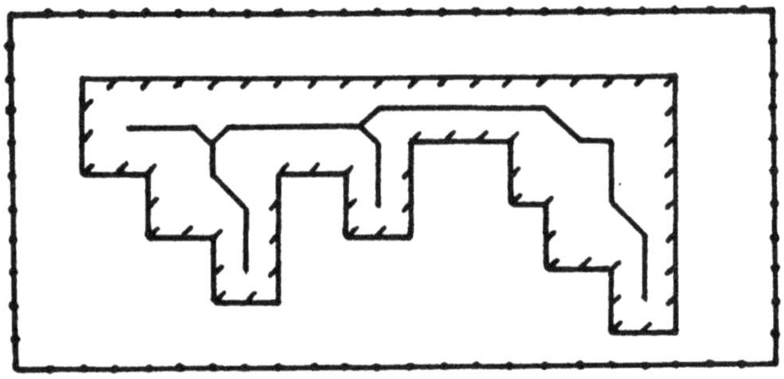

Fig. 4. The simple picture and its skeleton.

Description of the algorithm. Let us define corner line as horizontal line in the picture V, containing two or more its corners. The algorithm prosesses the picture using line-by-line scanning, so that the boundary of the scanned part successively occupies positions coinciding with the corner lines in this picture.

Let i be the number of the current corner line and y_i be its ordinate. The processing of the scanned part of the picture below this line results in detecting some segments of the skeleton and also forming so called residual picture $V_o(y_i) \subset V$ containing no concave corners of the types 5 and 8. The way to form the residual picture will be defined below. Let y_{i+1} be the ordinate of the next $(i+1)$ corner line. The scanned part $V_s(y_{i+1})$ of the picture V before processing of this line represents an union of the residual picture $V_o(y_i)$ and the part of the picture V, placed between i and $(i+1)$ corner lines.

The picture $V_s(y_{i+1})$ consists of one or some singly connected components V_K. Each component belongs to one of three classes: 1) class A of the simple pictures with upper boundary ordinate being equal to y_i; 2) class B of the simple pictures with upper boundary ordinate being equal to y_{i+1} and 3) class C of the pictures with upper boundary ordinate being equal to y_{i+1} and with contour containing one or more corners of the types 5 and 8 with the ordinate y_i.

The algorithm processes each component V_K as follows. If the picture belongs to the class A then succesive decomposition and skeleton detection are carried out in this picture. If the picture belongs to the class B then this picture represents the part of the residual picture and therefore remains unchanged. Finally, if the picture V_K belongs to the class C then for the first of its corners of the types 5 or 8 the dissective square is to be formed. Then the decomposition of the picture V_K into the left V_{1K} and the right V_{2K} parts is carried out.

The picture V_{1K} can belong to the classes A or B. In the first case the skeletonization of the picture V_{1K} is performed out and in the second case the picture is stored as the part of the residual picture $V_0(y_{i+1})$.

The picture V_{2K} can belong to the classes A, B or C. In the first two cases the executed steps are the same as considered above. In the third case the same steps previously applied to the initial picture V_K should be applied to the picture V_{2K}, i.e. the corner detection of the types 5 or 8 and so on.

The contour of the residual picture contains no corners of the types 5 and 8. So the following statement is valid.

Statement 3. The number of the corners in the residual picture does not exceed the value $2\cdot n$, where n is the picture line length. It follows that the main memory of the computer required for algorithm implementation does not depend on the length of the scanned part of the picture despite the size of determined maximal squares can be however large.

The algorithm has the most effective implementation when it uses an array of the coordinates of the corners $(x_1, y_1, x_2, y_2, \dots x_N, y_N)$ of the picture as the initial data. These coordinates are arranged in such order $i = 1, 2, \dots N$ that $i < j$ if $y_i < y_j$ or $y_i = y_j, x_i < x_j$. These data give the compressed coding of the picture ensuring its error-free restoration and therefore determination of the types of all its corners also. The algorithm processes the picture along its corner lines. Input coordinates of the corners along any current corner line actually set the boundary of the scanned part of the picture forward the level of the next corner line.

Perfomance. The algorithm has been implemented in the hardware-software system "Teremki" for automatic input, processing and semantic interpretation of the colour graphical pictures [3]. Despite the seeming complexity of the described algorithm the developed software implements it using only two-pass scanning of the residual picture contour for each corner line in the picture V. Processing speed of this algorithm implemented in the system is about $6\cdot 10^4$ pixels per second during coding of sufficiently complex pictures that is faster at least in several times in comparison with known algorithms. The processing speed is maximal for the pictures composed of horizontal and vertical lines and decreases for the pictures containing slant lines due to increasing number of the corners in these pictures.

Acknowledgments. The author is thankful to Dr. M.I.Shlesinger for the scientific supervision and some basic ideas.

References

[1] Stefanelli S., Rosenfeld A.: Some parallel thinning algorithms for digital pictures. JACM, 18 (1971), 255-264.

[2] Pavlidis T.: Athinning algorithm for discrete binary images. CGIP, 13(1980), 142-157.

[3] Macello V.V., Provalov Y.L., Sharipanov V.M., Shlesinger M.I.: The system for automatic input and processing of the graphical information. USiM, 6(1986), 56-60 (in Russian).

EXPERIMENTS WITH THE RECURSIVELY IMPLEMENTED CANNY EDGE DETECTOR

A. Lerch[1] and J. Fiser[2]

Introduction: Edge detection is a relevant tool of image processing. Its purpose is to transform a grey level image into an "edge image". A lot of algorithms have been developed to achieve better and better results. Edge detection suffers from the very same problem as most image processing tasks do, namely that different edge detectors can be found optimal for different classes of images. The goal of this pilot experimental study is to select a fast and reliable edge detector which can cope with rather bad quality pictures without loosing important features. A recent paper of Deriche [2] presents a new robust algorithm which is claimed to perform very well in noisy images. In order to evaluate the performance of this algorithm we chose the well-known Prewitt edge detector [3], which is one of the simplest proposed operators. We compared the edge images obtained by the two operators.

I. The Canny edge detector and its modification

One expects an edge operator to fulfill the following criteria:

1. It should respond to every valid edge point, i.e. there should be a low probability of missing real edge points.
2. It should not respond to false points which are introduced by noise even though they are local gradient maxima, i.e. there should be a low possibility of marking points falsely.
3. Points marked as edge points should coincide with real edge points, i.e. the average (or maximum) distance between the ideal and detected edge points should be small.

These criteria are treated by Canny [1] for a specific edge and noise model. He used a more or less acceptable assumption that the input image is degraded by additive Gaussian noise. The first two criteria result in investigation of the signal-to-noise ratio (SNR). The maximum achievable SNR corresponds to the best detection. The third criterion is the "localization" one, which involves the reciprocal of the standard

---

1) Computer and Automation Institute, 1111 Budapest, Kende u. 13-17., Hungary

2) Technical University of Budapest, Dept. of Electron Devices, 1111 Budapest, Goldmann Gy. tér 1-3., Hungary

deviation of the spatial distribution of detected edge points centered at the ideal edge. The examination of these three criteria yields an intuitively clear uncertainty principle between localization and detection, i.e. trying to better localize edges by using smaller operators results in weaker responses. Canny also introduced an additional constraint: the detector should not give multiple responses to one edge.

He used numerical optimization for determining the function class which maximizes the product of the detection and localization criteria under a constraint derived from the multiple response criterion. He approximated the optimal function with the first derivative of a Gaussian whose variance is a parameter of the filter. The 2D filter performs smoothing in edge direction by means of a Gaussian with the same σ. The broader this tunable filter is the better detection and the worse localization we obtain and vice versa.

Recently, Deriche [2] proposed an efficient realization and a small modification of Canny's filter. The Deriche filter is recursive, which means considerable savings in computational effort when applying larger filters. Only a small constant number of operations is necessary in each image point independently of the filter size. Using infinite support filter Deriche concludes that function

$$f(x) = c\, e^{-\alpha|x|} \sin\Omega x \qquad (1)$$

(in the gradient direction) is optimal. Here α and Ω are positive values and c is a constant.

We chose Deriche's algorithm because it has strong mathematical support and is optimal in the case of additive Gaussian noise. Also, its computational requirements are constant at different scales.

II. Investigation

When studying the features of the Deriche edge detector the following questions were investigated:

1. Tuning the filter is done by varying two parameters α and Ω. The behaviour of the filter was studied in the 2D parameter plane. We tried to separate the regions where the filter function oscillates and where one parameter dominates.

2. Several scale filters were tested in synthesized images and the results were compared to the results of the Prewitt operator. The issues

considered were detection and localization in the presence of synthesized noise.

3. The performances of the Prewitt and the Deriche filters were compared for bad quality low resolution real-world images. Typical indoor pictures and biological images were digitized into 128*128*6 bit matrices.

4. To avoid too subjective evaluation of the resulting gradient pictures a post-processing phase was added (simple thresholding or thresholding with hysteresis, and nonmaximum suppression). Automatic threshold selection seemed impossible thus thresholding was performed manually.

III. Results

1. According to Deriche [2], the best trade-off between localization and detection is achieved when the α/Ω ratio is large enough, although it is the worst case in the sense of the multiple response criterion. Our experiments seem to support Deriche's conclusion that sacrifying the multiple-response criterion does not mean any perceivable loss in the operator performance for bad quality pictures. In this case the single parameter α is sufficient for determining the filter behaviour, therefore tuning the operator becomes easier. The exponent in the filter function (1) suppresses the oscillation. We determined the region where the one-parameter-tunability is acceptable. Within this region we calculated the operator widths corresponding to different α values.

When α is comparable to Ω and the filter function oscillates, one can also achieve nearly as good results as in the previous case. However, we concluded that using this region is not advisable because small changes in the parameters can entail drastic changes in the resulting images.

2. Comparing the Prewitt and the Deriche filters <u>of similar size</u> for real-world imagery we experienced that

- the Deriche operator does not indicate more significant edges than the Prewitt operator
- the Deriche operator indicates fewer false edges due to noise; this effect was stronger in better quality images with higher contrast.

Increasing the operator size made the Deriche filter less noise sensitive as expected, but this suppressed a number of high frequency edges. The edge images became clearer, only the strong edges remained.

However, effects due to strong blurring, such as rounding-off and bad localization, became significant.

We conclude that in the case of relatively low resolution, low contrast pictures using the Deriche edge detector does not offer considerable advantage compared with the very simple Prewitt operator . The reason presumably is that low dynamics of pictures has stronger effect than noise. In certain applications the illumination of the scene can be similarly weak, therefore using simple hence computationally less demanding edge detectors should be considered.

In our investigations we did not exploit the advantages of multi-resolution edge detection. Selecting the "best-tuned" operator is quite difficult. Some approaches examine the behavior of the edge image in scale-space, i.e. continuously changing the operator size. Since Deriche's filtering is of constant complexity in scale-space, his filter is very favourable for such purposes.

References

[1] J.Canny: "A Computational Approach to Edge Detection", IEEE Transactions on Pattern Analysis and Machine Intelligence, Vol. PAMI-8, No. 6, Nov. 1986.

[2] R.Deriche: "Using Canny's Criteria to Derive a Recursively Implemented Optimal Edge Detector", International Journal of Computer Vision, pp. 167-187, 1987.

[3] J.M.S.Prewitt: "Object Enhancement and Extraction", in Picture Processing and Psychopictorics, B.S.Lipkin and A.Rosenfeld (Eds.), N.Y., Academic Press, 1970.

A NEW APPROACH FOR IDENTIFICATION AND ASSIMILATION OF POLYGONAL OBJECTS

N. Eichhorn, H. Kiesewetter

Academy of Sciences of the GDR
Central Institute for Cybernetics
and Information Processes
GDR 1086 Berlin, Kurstrasse 33

1. The identification problem

Let a polygonal object be defined by a non-self-intersecting cyclically closed polygonal contour.
Automatic recognition of polygonal objects requires the solution of the identification problem. Given a certain class of polygonal objects by the prescription of the ideal form of the contour, a class of admissable geometric transformations (e.g. translation, rotation, scaling) and an admissable deviation from the ideal contour. The problem is to decide, whether or not, a real polygonal object from a real scene belongs to the same class.
Many applications of digital image analysis (e.g. shape analysis, feature matching, visual inspection, quality assurance, selection of homologous structures in multispectral, multitemporal and stereo images) depend on the solution of this problem under different constraints.
The ideal object will be called invariant object, while the real object takes the role of the variable object, according to geometric transformation of the object as a whole and admissable deviation from the ideal form.
In general, overall coincidence cannot be achieved. Therefore, the question rises, if at least certain parts of the variable object can be identified with corresponding parts of the invariant object. The variable object has to be fitted to the invariant object in such a way that corresponding positions can be attached one to the other. The attempt to solve this problem offers a dialectic contradiction:

- Fitting of objects needs attached positions
- Attaching of positions needs fitted objects

To escape the dilemma, we propose an acceptable solution for an initial fitting: Transformation of both objects to a common eigen coordinate system.
Different modes of gravity (position, line, region) have to be considered for calculation of axial moments, gravity centres, eigenvalues and eigenvectors.
The attaching problem can easily be handled, if a reasonable fitting has been achieved. The basic information for attaching is provided independently of the actual value of the deviation parameter. This allows for fast variation of this parameter without repeating the basic calculation for attaching.
Fitting without attaching means that objects are identified by comparing its moments. These methods are well established in digital image analysis /1/. On the other hand, attaching without fitting commonly implies some apriori knowledge about the object. Our approach is based on the combination of fitting and attaching.
Let us suppose, that the problem of fitting and attaching can be solved in the above sense. Then we may ask for an assimilation of the variable object to the invariant object, at least at

those parts, where a certain correspondence can be confirmed. One form of assimilation will be, to substitute attachable parts of the variable object by corresponding parts of the invariant object. Repeated application of this type of assimilation represents some technical process of abstracting ideal forms out of real ones.

2. Fitting transformation

A position P with respect to a cartesian system of coordinates XP,YP will be denoted by

$$P = (XP, YP) \tag{1}$$

and two orthogonal vectors U and V by

$$U = (\ XU,\ YU) \tag{2}$$
$$V = (-YU,\ XU) \tag{3}$$

The axial moments, to be introduced, are related to a threedimensional rotation of the twodimensional polygonal object about the axis V. U indicates the orientation of the radius of rotation.

$$\|U\| = sqrt(XU*XU + YU*YU) \tag{4}$$

$$UP = (\ XU*XP + YU*YP) / \|U\| \tag{5}$$

$$VP = (-YU*XP + XU*YP) / \|U\| \tag{6}$$

$$F(M,U) = G * \int dP*UP**M \tag{7}$$

The zero moment is independent of U

$$F(0,U) = G * \int dP = F0 \tag{8}$$

Let U be a fixed direction and consider the second moment

$$\int dP*(UP-UC)**2 \tag{9}$$

in dependence of translation UC of the axis V. For arbitrary U, the optimum value of UC is achieved at the centre of gravity.

$$UC = F(1,U) / F0 \tag{10}$$

$$F(1,U) = G * \int dP*(XU*XP + YU*YP) / \|U\| \tag{11}$$

$$UC = (XU*XC + YU*YC) / \|U\| \tag{12}$$

$$XC = \int dP*XP / (\int dP) \tag{13}$$

$$YC = \int dP*YP / (\int dP) \tag{14}$$

The optimum orientation U with respect to (9) results from the eigenvalue problem

$$UU*XU + UV*YU = L*XU \tag{15}$$

$$UV*XU + VV*YU = L*YU \tag{16}$$

$$UU = G * ((\int dP)*(\int dP*XP*XP) - (\int dP*XP)*(\int dP*XP)) \tag{17}$$

$$UV = G * ((\int dP)*(\int dP*XP*YP) - (\int dP*XP)*(\int dP*YP)) \quad (18)$$

$$VV = G * ((\int dP)*(\int dP*YP*YP) - (\int dP*YP)*(\int dP*YP)) \quad (19)$$

Let UL be the maximum and VL the minimum eigenvalue L and

$$RR = sqrt(4*UV**2 + (UU-VV)**2) \quad (20)$$

Then we get the following formulas:

$$UL = (UU + VV + RR) / 2 \quad (21)$$

$$VL = (UU + VV - RR) / 2 \quad (22)$$

For RR = 0 no eigenvector can be specified (rotational symmetric object).

For UU $\geq$ VV :

$$XU = 1 \quad (23)$$

$$YU = 2*UV / (RR + abs(UU-VV))$$

For VV > UU :

$$XU = 2*UV / (RR + abs(UU-VV)) \quad (24)$$

$$YU = 1$$

The general orthogonal transformation to the eigen coordinate system takes the form

$$X = (XU*(XP-XC) + YU*(YP-YC)) / (\|U\| * UL) \quad (25)$$

$$Y = (-YU*(XP-XC) + XU*(YP-YC)) / (\|U\| * UL) \quad (26)$$

It includes scaling, rotation and translation as special cases. Substituting

$$UL = 1 \quad (27)$$

means no scaling and

$$(XU, YU) = (1,0) \quad (28)$$

means no rotation.
The fitting transformation of the variable object (X1, Y1) to the invariant object (X2, Y2) is derived from the condition that the eigen coordinate systems coincide.

$$(X1, Y1) = (X2, Y2) \quad (29)$$

The centre of gravity ((13), (14)) and the eigen matrix ((17), (18), (19)) have to be calculated for different mode of gravity. We propose a consistent calculation scheme for position, line and region mode. Since integration can be carried out explicitly for polygonal objects, we get formulas with infinitesimal accuracy in terms of the coordinates /2/. Summation runs over positions, resp. line segments, resp. triangular regions.
For line mode we introduce the euclidean length of a line segment from INIT (XI, YI) to EXIT (XE, YE).

$$LEN = sqrt((XE-XI)**2 + (YE-YI)**2) \qquad (30)$$

The corresponding quantity for region mode proves to be the double area of the triangle origin_INIT_EXIT.

$$DAR = XE*YI - YE*XI \qquad (31)$$

The double area may assume positive or negative values in dependence of positive or negative orientation of the contour. The region mode formulas can only be applied for non-self-intersecting closed polygons ($TT \neq 0$).

$$\begin{aligned} TT &= \sum (1) && (32)\\ &= \sum LEN \\ &= \sum DAR \end{aligned}$$

$$\begin{aligned} XT &= \sum XP && (33)\\ &= \sum LEN*(XI+XE) \\ &= \sum DAR*(XI+XE) \end{aligned}$$

$$\begin{aligned} YT &= \sum YP && (34)\\ &= \sum LEN*(YI+YE) \\ &= \sum DAR*(YI+YE) \end{aligned}$$

$$\begin{aligned} XX &= \sum XP*XP && (35)\\ &= \sum LEN*(XI*XI + XE*XE + (XI+XE)*(XI+XE)) \\ &= \sum DAR*(XI*XI + XE*XE + (XI+XE)*(XI+XE)) \end{aligned}$$

$$\begin{aligned} XY &= \sum XP*YP && (36)\\ &= \sum LEN*(XI*YI + XE*YE + (XI+XE)*(YI+YE)) \\ &= \sum DAR*(XI*YI + XE*YE + (XI+XE)*(YI+YE)) \end{aligned}$$

$$\begin{aligned} YY &= \sum YP*YP && (37)\\ &= \sum LEN*(YI*YI + YE*YE + (YI+YE)*(YI+YE)) \\ &= \sum DAR*(YI*YI + YE*YE + (YI+YE)*(YI+YE)) \end{aligned}$$

$$\begin{aligned} XC &= XT/TT && (38)\\ &= XT/(TT*2) \\ &= XT/(TT*3) \end{aligned}$$

$$\begin{aligned} YC &= YT/TT && (39)\\ &= YT/(TT*2) \\ &= YT/(TT*3) \end{aligned}$$

$$\begin{aligned} UU &= TT*XX - XT*XT && (40)\\ &= TT*XX*2 - XT*XT*3 \\ &= TT*XX*3 - XT*XT*4 \end{aligned}$$

$$\begin{aligned} UV &= TT*XY - XT*YT && (41)\\ &= TT*XY*2 - XT*YT*3 \\ &= TT*XY*3 - XT*YT*4 \end{aligned}$$

$$\begin{aligned} VV &= TT*YY - YT*YT && (42)\\ &= TT*YY*2 - YT*YT*3 \\ &= TT*YY*3 - YT*YT*4 \end{aligned}$$

3. Attaching of positions

For arbitrary position P1 of the variable object, we construct an attached position P2 of minimum distance at the invariant object.

Let

$$U = E - I \tag{43}$$

be an arbitrary chord of the invariant object from I to E and V the normal vector according to (3). We define the quadratic dipole distance by the formula

$$QIE(P1,I,E) = \tag{44}$$

$$\|P1 - I\|^{**}2 \qquad \text{for } (P1-I,U) \leq 0$$

$$\|P1 - E\|^{**}2 \qquad \text{for } (P1-I,U)^{**}2 \geq \|U\|^{**}2$$

$$(P1-I,V)^{**}2 / \|U\|^{**}2 \qquad \text{else}$$

and the quadratic deviation by

$$DEQ(P1) = \min QIE(P1,I,E) \tag{45}$$

Formula (44) uses the common notation of norm by $\| \ \|$ and scalar product by (,).
The minimum has to be taken over all chords of the invariant object.
The minimum value is at least attained at one chord. The nearest position P2 at such a chord is called an attached position of P1 at the invariant object. It may be the INIT or EXIT position or an inner position of this chord.
Given an admissable deviation DEV, a variable object is decided to belong to the same class as the invariant object, if deviation condition

$$DEQ(P1) \leq DEV*DEV \tag{46}$$

holds for each position P1 at the contour of the variable object.

4. Assimilation to the ideal object

In general, only certain segments of the variable object will fulfil deviation condition (46). Let I1 be the INIT, E1 the EXIT of such a segment and I2 resp. E2 corresponding attached positions at the invariant object. Then this segment will be replaced by the straight line segment from I1 to I2, the corresponding segment at the contour of the invariant object from I2 to E2 and the straight line segment from E2 to E1.

5. Literature

/1/ Qiuming Zhu; Lay-kheng Poh: A transformation-invariant recursive subdivision method for shape analysis, 9th international conference on pattern recognition, Rome, Italy, 1988, pp.833-835

/2/ Eichhorn,N.; Kiesewetter,H.: INTEGER analysis of two-dimensional polygonal objects, Computing 1989, in press

COMPLEX MOMENTS DESCRIPTOR OF CONTOUR SEQUENCE II

A. Abo Zaid , E. Horne

Electronic Engineering Labs.,University of Kent, Canterbury,Kent,U.K. , CT5 7NT

ABSTRACT

This paper proposes a novel approach for using boundary moments in pattern recognition , in which the complex moments descriptor of the one dimensional complex function representation of the contour is the feature vector . Experimental results presented show that the method has better performance characteristics in comparison with other methods .

1. INTRODUCTION :

A careful examination of the boundary of a closed planer shape reveals that similar shapes have similar boundaries . That is , essential information is contained in the boundary of the shape . In order to use this information as a basis for shape recognition , an appropriate model for the boundary is required . Shape descriptors based on boundary representation have two sub-problems . The first of these problems is related to methods of representing the image boundary function without loss of information content . Most of the techniques usually used in shape recognition expresses the boundary in terms of the magnitude of radii or the phase angles . Such representations involve a loss of information and as a result , whatever the properties of the transform used , the output will have some inefficiency. The second problem lies in the transform used . However , the method of moments furnishes a good systematic procedure for extraction of a set of quantitative features from an image .

The contour of a closed planer digitised object can be describe by an order pair sequence : Z { x(i) , y(i) } = { z(1) ,z(2) , ...,...,z(n) } of n equi-spaced points on the boundary of the contour considered . If the origin of the measurement axis is taken to be the centeroid of the object then , the contour sequence $Z_t(i)$ of the translated shape is identical to the contour sequence Z(i) of the original shape . If a shape is uniformly scaled by a factor λ , the area of the polygonal shape represented by the contour is scaled by λ^2 . Therefore the length of the boundary and the vectors connecting the centroid and the boundary are scaled by λ . As a result , the duration as well as the amplitude of the contour sequence are scaled proportionately . Hence , the contour sequence representation of the scaled shape has the same profile as that of the original shape . Moreover , if only the magnitude of the sequence Z(i) is considered , the resulting sequence is rotationally invariant .

Although moments of contour have been used for extracting features in pattern recognition , there is less information related to this approach in published literature . Further , some references indicate that the feature vector is formed by the same method as that used for moments of silhouette [5] . In consequence it is found that a significant reduction of higher order moments information occurs since the results are dominated by low order moments .In subsequent work [8] , it was shown that the technique required for normalising contour data with respect to scale is different from that required for silhouette data .

Gupta [6] has used the moments of the magnitude of the function Z(x(i),y(i)) , above , for shape recognition . Although the approach is simple and efficient when the distance between the object classes considered is large , its performance deteriorates when the classes under investigation are close . In a subsequent work [7] , he considered the moments of the ordered pair sequence above in the x-y plane . Since the moments generated , in this case , are not invariant if the object undergoes rotation , a pre-alignment of the contour in a fixed orientation in space or dynamic alignment are used . While the first approach is subject to noise [7] , dynamic alignment is computationally costly .

2. COMPLEX MOMENTS DESCRIPTOR :

Complex moments are a recent approach for deriving moment invariants . They form an intermediate step between ordinary moments and moments invariants [3,4] . The complex moments of order (p,q) for a two dimensional function f(x,y) are given by :

$$C_{pq} = \int_{-\infty}^{\infty}\int_{-\infty}^{\infty} (x+jy)^p (x-jy)^q f(x,y)\,dx\,dy$$

If f(x,y) is real non-negative , the contrast of the image , C_{pp} is a real non-negative number , while C_{pq} is the complex conjugate of C_{qp} . The complex moment of order (p,q) is a linear combination of the ordinary moments M_{rs} with the relation :

$$r+s=p+q=m$$

The complex moments descriptors (CMD) [1] , are both very simple and quite powerful in providing an analytic characterisation for moment invariants . They differ from complex moments in the method of normalisation and thus the nature and efficiency of the transform output produced . In the CMD the amplitude of each coefficient is equal to the amplitude of the corresponding complex moment and , the phase is a linear combination of the phase of the corresponding complex moments coefficient and the phase of a selected reference component . Two algorithms for calculating the phase of the CMD coefficients can be used . In the first , the reference component is selected from the largest coefficients whose angular order , (p-q) , is equal to one i.e. C_{21} , C_{32} ,...,*etc.* . In this case , the reference component's phase expresses the rotation angle of the object . Thus if (p-q) times this phase is mixed with the phase of the complex moments component of order pq , the result , which is the phase of the CMD component of order pq , will be independent of the object orientation in space . Moreover , in this case only , i.e. when the angular order of the reference component used is unity , the CMD is the complex moment when the image has a fixed orientation in space . However , it may be that all the coefficients of the complex moments with unity angular order are of small amplitude and their use in the normalisation process , in the presence of noise , can produce unsatisfactory results . In this case the generalised complex moments descriptor approach can be used , where the reference component is the coefficient with the maximum complex moments amplitude and angular order greater than zero . The combination process and the physical interpretation of the CMD is different . The CMD in the generalised approach is just a mathematical tool to form a feature vector which is independent of object transformation and has a good noise immunity . However , the phase angle of the CMD in this case is m times (m is the angular order of the reference component used) that of the corresponding complex moments at a fixed orientation in space . Despite this classification in this case is quite simple [2] . For both cases , the calculation required is simple and the output is a moments feature vector which is invariant to object transformation .

3. COMPLEX MOMENTS DESCRIPTOR OF CONTOUR BOUNDARY :

definition :

The complex moment of contour C, CM^B_{pq}, is defined by :

$$CM^B_{pq} = \int_{x,y \in C} (x+jy)^p (x-jy)^q \, dx dy \tag{1}$$

Given the contour of a binary image in a sequence $Z(i,x,y) = x(i) + jy(i)$,i=0,1,.....,N-1 , with x,y measured from the centeroid of the shape, the complex moment of contour CM^B_{pq} is :

$$CM^B_{pq} = \sum_{i=0}^{N-1} [x(i)+jy(i)]^p [x(i)-jy(i)]^q \tag{2}$$

and :

$$CM^B_{11} = \sum_{i=0}^{N-1} x^2(i)+y^2(i)$$

The complex moments of a contour boundary can be calculated from equation 2 directly using complex multiplication or from their relation to the ordinary moments (figure 1-a) and a hardware structure as shown in figure 1-b. It is clear from figure 1-b that the ordered pair sequence (x,y) form the input and the CMD are calculated as a weighted sum of the ordinary moments with the same order.

Normalisation :

[1] Translation : The contour sequence $Z(i,x,y) = x(i) + jy(i)$ with x,y measured from the centeroid is invariant with respect to object translation.

[2] Scale : When the contour undergoes dilation, the new complex contour moment CM'^B_{pq} is :

$$CM'^B_{pq} = \lambda^{(p+q)} CM^B_{pq} \text{ and :} \tag{3}$$

$$CM'^B_{11} = \lambda^2 CM^B_{11}$$

Thus, dilation normalisation is achieved by :

$$CM^B_{pq} = CM^B_{pq} \cdot \left(\frac{N}{CM^B_{11}}\right)^{\frac{p+q}{2}} \tag{4}$$

where N is the number of points considered.

[3] **Rotation :** When the object rotates by angle θ, each point Z(i,x,y) on the contour sequence is transformed such that $Z' = Z\, e^{j\theta}$. From (2) it is clear that :

$$CM'^B_{pq} = CM^B_{pq} \cdot e^{j(p-q)\theta} \tag{5}$$

In order to normalise the complex moments of the contour sequence with respect to rotation, the principle of the complex moments descriptor discussed earlier can be used. Thus, the amplitude of the calculated moments is retained, while the argument of each moment coefficient of order pq, α_{pq}, is replaced by a new angle given by :

$$\phi_{pq} = \alpha_{pq} - (p-q)\alpha_* \qquad \textit{where :} \tag{6}$$

α_* is the phase angle of CM'^B_{32} *or* CM'^B_{21}. The resulting phase of each CMD component will be invariant to object rotation. Moreover, in this case, the CMD is in fact, the complex moments of contour sequence at a specific orientation in space independent of the object rotation.

As previously discussed, some shapes may have coefficients whose angular order is unity with a small amplitude which makes them sensitive to noise. In this case the generalised CMD approach should be used. The algorithm for generating the CMD of a contour feature vector can be summerised as follows:

[1] Normalise for size by dividing CM^B_{pq} by $(\frac{CM^B_{11}}{N})^{\frac{p+q}{2}}$.

[2] If the largest magnitude of the complex moments is $CM^B_{p_m q_m}$, $(p_m \neq q_m)$, find the phase of CMD coefficient of order pq, ϕ_{pq}, from the following relation:

$$\phi^{ref}_{pq} = arg\,[(p_m - q_m)\alpha^{ref}_{pq} + (p-q)\alpha^{ref}_{p_m q_m}] \tag{7}$$

where α^{ref}_{pq} is the argument of CM^B_{pq} of the reference contour. Store $|CM^B_{pq}|$ and ϕ^{ref}_{pq} as the normalised complex moments descriptor of the reference contour.

[3] For the sample contour C', repeat steps [1-2], Store $|CM'^B_{pq}|$ and ϕ^{samp}_{pq} as the normalised complex moments of the sample contour.

From the above it is clear that; $\phi^{ref}_{pq} - \phi^{samp}_{pq} = (p_m - q_m)\,[\alpha^{ref}_{pq} - \alpha^{samp}_{pq}]$ and this value is independent of rotation of the shape [2].

4. DISTANCE MEASURE AND EXPERIMENTAL RESULTS:

The normalised complex moments descriptor has a phase difference angle m times the real phase difference angle of the complex moments coefficients, $m = p_m - q_m$, thus the error angle ε^d_{pq} is given by:

$$\varepsilon^d_{pq} = \frac{\phi^{ref}_{pq} - \phi^{samp}_{pq}}{p_m - q_m} \tag{9}$$

and the minimum distance between the sample and reference contour, d, is given by:

$$d = \sqrt{\sum_{p,q} |CM^B_{pq}|^2 + |CM'^B_{pq}|^2 - 2|CM^B_{pq}|.|CM^B_{pq}|\cos\varepsilon^d_{pq}} \tag{10}$$

It is clear that, if the sample contour is the same as the reference contour then $\alpha^{ref}_{pq} = \alpha^{samp}_{pq}$ and equally is $\phi^{ref}_{pq} = \phi^{samp}_{pq}$. As a result, the phase error, ε^d_{pq}, is usually small (ideally zero) if the two contours belong to the same class.

Two experiments are used to test the algorithm in comparison with the method which uses the moments of the amplitude of contour sequence [6]. In the first experiment, six classes of objects (Figure 2-a) were used to test the algorithm. For each object model 2400 samples are generated with varying position, scale and, rotation for different levels of noise up to -24.5 db. Noise represented by a normal distribution in x and y (noise level: -40 db to -24.5 db of minimum enclosed rectangle) was added to each test contour. Classification is based on on a minimum distance classifier in which the unknown sample represented by the CMDs is assigned to class K^* such that

$$K^* = arg\,(\min_k \; D[k])\;, k=0,1,2,3,4,5$$

where D[k] is the Euclidean distance between the test vector and the reference vector given by the norm calculated

between the two vectors. Figure 2-b shows six test samples three belonging to class 2 and the other three belonging to class 3 and figure 2-c the classification rate for both methods.

In the second experiment , four 64 by 64 pixel real world image models (figure 2-a) were used . The noise model in this experiment is defined as follows [6] : the probability that each contour pixel will move to one of its 8NNH is taken to be β=0.7 , this produces "noise level 1" (table 2.1.1 , 2.2.1) . A repetition of this process give "noise level 2" (table 2.1.2 , 2.2.1) . The images are also rotated . For each model , 100 such images were used in testing as before . The resulting confusion matrix is given in table 2 and figure 3.b shows a sample from the test images for each model .

The second experiment was repeated using real images of shape 2,3 from experiment one only . The total classification rate when only the magnitude of the contour sequence is used is 82 % in comparison to 90.5 % for the proposed method . It may be worth noting that the classification rate for the Fourier descriptor method in a similar experiment was found 92.5 % using the optimum normalisation method [2] .

5. CONCLUSION :

The principle of complex moments descriptor is extended to present a novel image feature vector . The new feature vector utilises the complex contour sequence representation , thus no information is lost . Moreover , it uses the complex moments transformation which features orthogonality of basis vector and uniqueness , in addition to good convergence and invariance characteristics . Experimental results show that the approach has a better classification rate in comparison with other methods .

6. REFERENCES :

[1] A. Abo-Zaid , O.Hinton , E. Horne , "About Moment Normalisation And Complex Moment Descriptor" , proceedings of the 4th International Pattern Recognition Conference , Cambridge , U. K. , pp. 399:409 , 28-30 march 1988 .

[2] A. Abo-Zaid , E. Horne , "Algorithm for normalisation of Fourier descriptor" , Proceeding of Visual Communications and Image Processing '88 , Cambridge , MA , USA (SPIE 1001 , PP. 599 :606) , 9-11 Nov. 1988 .

[3] Y. Abu-Mostafa and D. Psaltis , "Recognitive Aspects of Moment Invariants" , IEEE T-PAMI , Vol. 6 , No. 6 , PP. 698:706 , 1984

[4] Y. Abu-Mostafa and D. Psaltis , "Image Normalisation By Complex Moments" , IEEE T-PAMI , Vol. 7 , No. 1 , pp. 46:55 , 1985 .

[5] S.A. Dudani , " Aircraft Identification by Moments Invariants ," IEEE T-Computer , pp. 171:176 , vol 26 , 1977 .

[6] L. Gupta & M.D Srinath , " Contour Sequence Moments For The Classification of The Closed Planer Shape " , Pattern Recognition , vol 20 , no 3 , pp. 267:272 , 1987 .

[7] L. Gupta & M.D Srinath , " Invariant Planer Shape Recognition Using Dynamic Alignment " , pattern Recognition , vol 21 , no. 3 , pp. 235:239 , 1988 .

[8] T.A. Grogan , " Shape Recognition and Description : Comparative Study " , Ph.D. Thesis , Purdue University , 1983 .

Figure 2-a

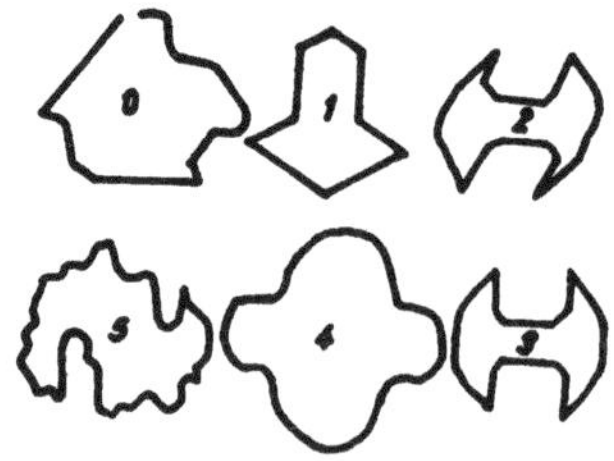

Figure 1-a : Relation between complex moments and odinary moments

[1] $Re\ C_{20} = m_{20} - m_{02}$
$Im\ C_{20} = 2m_{11}$

[2] $Re\ C_{11} = m_{20} + m_{02}$
$Im\ C_{11} = 0$

[3] $Re\ C_{30} = m_{30} - 3m_{12}$
$Im\ C_{30} = 3m_{21} - m_{21}$

[4] $Re\ C21 = m_{30} + m_{12}$
$Im\ C21 = m_{21} + m_{03}$

[5] $Re\ C40 = m_{40} - 6m_{22}$
$Im\ C40 = 4m_{31} - 4m_{13}$

[6] $Re\ C_{31} = m_{40} - m_{04}$
$Im\ C_{31} = 2m_{13} + 2m_{13}$

[7] $Re\ C_{22} = m_{40} + 2m_{22} + m_{04}$
$Im\ C_{22} = 0$

[8] $Re\ C_{50} = m_{50} - 6m_{32} + 4m_{14}$
$Im\ C_{50} = 4m_{41} - 6m_{23} + m_{05}$

[9] $Re\ C_{41} = m_{50} - 6m_{32} + 4m_{14}$
$Im\ C_{41} = 3m_{41} + 2m_{23} - m_{05}$

[10] $Re\ C_{32} = m_{50} + m_{32} + m_{14}$
$Im\ C_{32} = m_{41} + 2m_{23} + m_{05}$

Figure 2-b

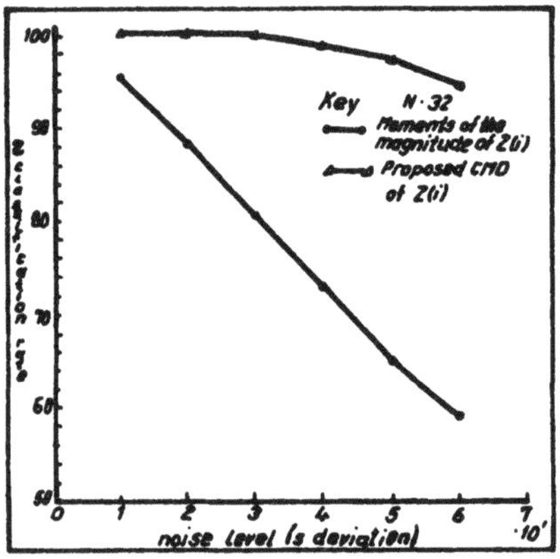

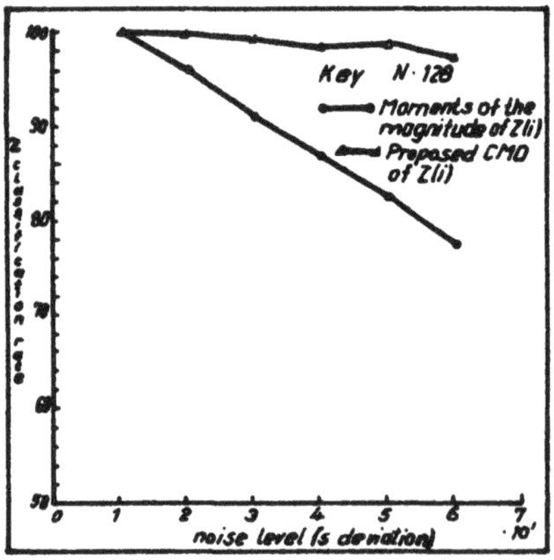

Figure 2-c

Figure 3-a

figure 3-b

Table (2)

1 - Moments of the magnitude of Z (i)

1.1 Noise Level (1)

ip-op	shape1	shape2	shape3	shape4
shape1	100	0	0	0
shape2	0	74	11	15
shape3	0	6	86	8
shape4	0	0	4	96

1.2 Noise level (2)

ip-op	shape1	shape2	shape3	shape4
shape1	96	0	3	1
shape2	2	61	18	19
shape3	0	9	75	16
shape4	1	1	7	91

2 - Proposed CMD of Z (i)

2.1 Noise level (1)

ip-op	shape1	shape2	shape3	shape4
shape1	94	0	0	6
shape2	0	100	0	0
shape3	0	6	94	0
shape4	0	0	0	100

2.2 Noise level (2)

ip-op	shape1	shape2	shape3	shape4
shape1	88	0	1	11
shape2	0	94	0	6
shape3	0	7	93	0
shape4	0	0	1	99

Figure 1-b: A scheme for claculating complex moments descriptor up 5th order

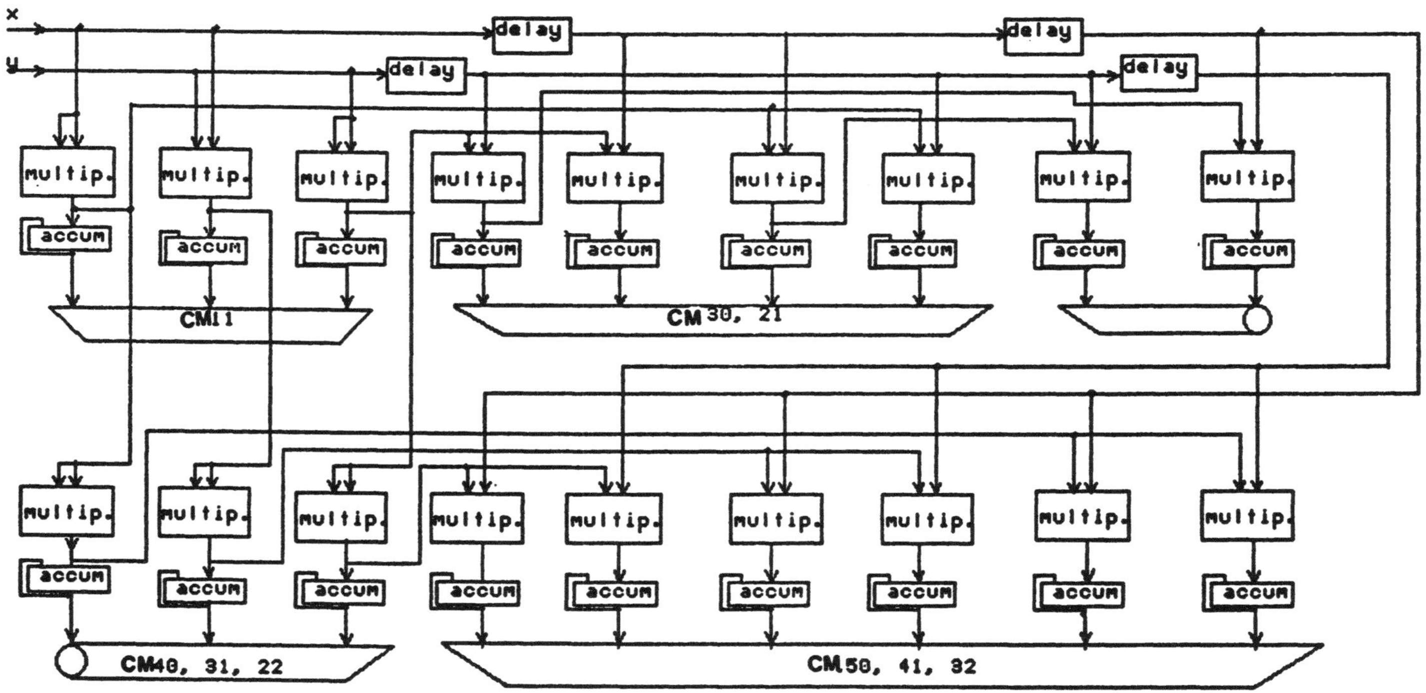

THE ATLAS-90 INTERACTIVE IMAGE PROCESSING DISPLAY SYSTEM

István Rényi[1], Sándor Zöld[1], Attila Fekete[2]

1. INTRODUCTION

Advances in the design of image processing display equipment have occurred in two key areas, namely, data flow architec-tures and pipeline array processor architectures. *Data flow* involves the use of high speed data interfaces for the displayprocessor, a flexible internal bus architecture and efficient access to display processor memories. Image processing power is achieved through the use of *pipeline processors* rather than high speed, large address-space CPUs. The key idea is that the value of the display in processing applications is its fast and convenient access to a large data base. This ease of access has helped the growth in processing applications, and the fact that in many cases the reduced precision used by the display is adequate for analysis by a trained observer.

The ATLAS-90 Interactive Image Processing System, developed by a joint team from Híradástechnika Szövetkezet (HT) and the Central Research Institute for Physics (KFKI), exemplifies the above architectural advances. In the first place the system is intended for processing and displaying *remotely sensed multispectral or multitemporal imagery*, but other application areas requiring fast processing and sophisticated display may also take advantage of its special features.

In order to satisfy the potentially broad base of users, each usually with a slightly different set of requirements, the following *objectives* were adopted in the design of the ATLAS-90:

- To provide a flexible data base to accomodate varying image size, precision and operational modes.
- To provide a powerful facility for interchanging data between the image memory and the processing elements and between the image memory and the external data sources.
- To provide a powerful basic processing capability and the possibility of including specialized processors for specific tasks.

The two key issues determining the ability of the harware to meet these objectives are a) Memory Architectures and b) Processing Power. Following a brief overview of the basic system components, the main features of the image memory architecture will be described along with the means and methods to improve image processing and display operations. Furthermore we

[1] Central Research Institute for Physics, 1525 Budapest, P.O.Box 49.
[2] Híradástechnika Szövetkezet, 1529 Budapest, P.O.Box 268.

give a global overview of the organisation of the software components. To show the present and potential effectiveness of ATLAS-90, a few examples of specific processing tasks will also be given.

2. HADWARE ARCHITECTURE

The basic components of ATLAS-90 are the following:

- Controlling minicomputer with coprocessor and conventional peripheral devices (various disks, tape drives, hardcopy and up to eight display terminals). The controller is a single board, PDP11 44 compatible processor (IPC) with a 64-bit Floating Point Processor (FPP) extension, both user microprogrammable and manufactured by KFKI.
- Image Memory Subsystem (see later).
- Pipeline Processor Array (see later).
- Display Subsystem. The processed image can be displayed in the RGB or the IHS (Intensity-Hue-Saturation) system in 512x512 format, 24 bits per pixel. Controlled by the colour assignment memory the processed image can be mixed with the output of the
  - -- graphic memory. A vector generator and four bitplanes are available which can be zoomed and scrolled,
  - -- symbol generator, producing 32x80 coloured characters,
  - -- two programmable cursor generators.
- Specialized processors (see later).
- Precision image digitizer. This - optional - equipment is used for inputting image data from various kinds of plain media, e.g. photoprints, films and maps with high precision.

Memory architecture is the foundation on which all of the capabilities of the system are built. Fast and efficient data routing is the goal of the memory design, in support of processing, display and data base operations. To this end, the image memory modules have nine, byte-wide, video speed (75ns/byte) ports: two times four of these are bussed to establish memory input and output lines. The fifth output port of each module directly feeds the pipeline processors. Up to 16 modules are incorporated in a system, each holding two blocks of 512x512 bytes. (Note that the access bandwidth for the entire 8Mbyte image data base is 210 Mbyte/s!) Specialized processors and external equipment access the image memory via the eight I/O buses, the maximum bandwidth here exceeds 100Mbyte/s.

The memory subsystem is controlled by a microprogrammed bipolar microprocessor (MC) which, in addition to orchestrating the array under various operation modes provides "hardware" zoom (2,3,...,16), scroll/pan and split screenfacilities and performs rigorous test sequences, including signature analysis on each individual module simultaneously.

Processing power. The key element to the processing capabilities of ATLAS-90 is the *pipeline processing channel.* Using the three pipeline processing channels one can generate a "virtual image" from a linear combination of all input channels. A large number of algorithms can be performed using the input and output function tables, adders, range circuits, etc. of the pipeline processors. A few examples are given at the end of the paper.

The computation of an image histogram is necessary in a wide variety of algorithms. The histogram cannot be obtained by the host since the image to be histogrammed is frequently a virtual image formed in the pipeline processor. In the ATLAS-90 a 20-bit histogram is computed by the so-called *Histometer* in two frame times per channel. For the computation, a Region Of Interest (ROI) memory is used to mask the image. The Histometer also measures the area of the ROI and the sum of all pixel values within the ROI.

The introduction of the *Feedback Processor* concept provides a significant advance in the processing capabilities. Previously all video speed processing was limited to a single iteration. With the feed-back loop the user may store the image as seen on the monitor back into refresh memory which could then be returned to the host processor or used in subsequent processing steps. The Feedback Processor includes tables, comparators, adders, a multiplier-accumulator and shifter, among others. Application examples will be given later.

The "hardware processing" capabilities of ATLAS-90 can be further extended by adding *specialized processor modules.* There are two architectural features of the system which make the effective use of such processors possible:

- the high memory bandwidth achievable via the multioperand memory I/O buses
- the flexible accessing capability of the image memory.

Until now no spec. processor module has been completed. From the first ones considered an image warper could be mentioned. This would produce geometric transformation of up to 1kx1k images using 1.5 degree polynomials with 1/8 pixel accuracy and nearest neighbour or bilinear interpolation. The processing time for a 512x512 image using bilinear interpolation is approx. 1.3s.

3. SOFTWARE ARCHITECTURE

The software system is built on the UNIX (version 7.), its components can be grouped as follows:

- the User Interface,
- the Image Processing Module Library,
- the Function Library,
- the UNIX kernel extension, activating the microprogrammed Image Memory Controller.

The images are stored on three different storage media (in several forms): on magnetic tapes, on disks, and in the Image Memory. A common characteristic of all these media is that an image always consists of two main components, a descriptor and the image data. The descriptor in each case contains information necessary for using and interpreting the image data . If image files are stored on disk or magnetic tape the descriptor contains information about files related to a given image, such as history, statistics or colouring information, etc.

The system has its own User Interface, which is of dual nature, as it has a menu oriented and a command language oriented component. Both the menu and command interfaces are partitioned into an application independent program and into a database containing all application dependent information. Databases of the user interface can be defined in a simple language; the database itself is generated by a translator program. Modification and extension of an existing system can be carried out by modifying the definition and regenerating the appropriate component.

The Menu System enables the user to select actions and parameter values using full screen menus, as well as to define parameters interactively. The Menu System checks the received parameter values against ranges and substitutes default values if necessary. There are further services available for the convenience of the user, e.g. history accumulation, inline command execution, etc.

The ATLAS-90 Command Language facility enables the user to work more efficiently and it ensures an interface compatible with that of the UNIX system. The realized Command Language is more user friendly, because it allows using of keyword and positional parameters; option and parameter identification by mnemonics is possible.

The ATLAS-90 Module Library is a set of user programs, which has to be designed according to some conventions. The program parameter interface (options, optional parameters, parameters) has to meet the requirements of the Menu System and Command Interpreter. Furthermore, the programs can access the Image Memory or activate any image processing services only with the help of the ATLAS Function Library. The first functional group of the Module Library is that of the utility programs:

- Image File Handling: the purpose of these modules is transferring image files between the system and the external world, or within the system: between different storage media, with the necessary administration.

- Screen Display Control: these modules are responsible for screen definition, assignment of image memory areas to subscreens, changing parameters (scroll, zoom), grey level control, split screen, etc.

- Test Programs: the Image Memory System has its own diagnostic programs, including signature analysis.

The second group consists of the image processing modules. Our intention is to provide a wide variety of these , although the main area of our interest is the processing of multispectral images. Some interesting examples of the available modules are as follows:

- Preprocessing of images (filtering, smoothing, corrections, etc),
- Image statistics,
- Geometric transformations,
- Fourier and inverse Fourier transforms (partially implemented in microcode),
- Supervised classification,
- Clustering with ISODATA algorithm,
- Image arithmetics: the program evaluates simple arithmetic expressions, where the operands are blocks of the Image Memory.

The Function Library of the system contains the following functional groups:

- Image Memory Access: the functions move data blocks between the Image Memory and the operative store of the host computer, handle the contents of the Image Memory through windows for processing purposes, etc.

- Logical Screen Control: these functions serve for defining subscreens, assigning Image Memory areas, scrolling or zooming displayed data , etc.

- Display Subsystem Control: these functions handle the complex facilities of the Display Subsystem,of the Pipeline Processor, of the Image Memory Controller, as well the interactive devices.

- Image File System Functions: this group provides elementary functions for image file handling.

The Image Memory Controller, a microprogrammed special processor is attached to the IPC controlling computer. Its physical handling is realized by a device driver included into the UNIX kernel.

4. APPLICATION EXAMPLES

The following list of operations gives a few examples of the ATLAS-90's capabilities, but does not include all of the power inherent in the system.

1) Addition/subtraction, position invariant intensity mapping: The PL can be used to perform weighted linear combination of images, gamma correction, contrast enhancement, level slicing, contouring or any other radiometric operation on monochrome or colour images in one frame time.

2) Convolution: Using the Feedback Processor (FP) path any NxM kernel convolution can be performed in NxM frame times.

3) Median Filtering: Applying the sorting module of the FP, 2-D median filtering can be performed using any shape of 9-point window in 1.6s.

4) Antivignetting: This operation is carried out in one frame time via the FP path, using the following equation:

$$Out(x,y) = In(x,y) * Weight(x,y)$$

5) Coordinate Transformation in 3-D space: 256K homogeneous coordinate points may be stored using the image warper and 8 image memories. All points are multiplied by a 4x4 matrix in 32 frame times:

$$[x',y',z',w'] = [x,y,z,w] \begin{bmatrix} a1 & b1 & c1 & d1 \\ a2 & b2 & c2 & d2 \\ a3 & b3 & c3 & d3 \\ a4 & b4 & c4 & d4 \end{bmatrix}$$

If only the FP is available for this operation the processing time is increased by a factor of 4.

6) Box Classification: For this simple and fast pixel classification method the FP can be applied if the dimensionality is ≤ 4. Processing time is $2c+1$ frame times, where c is the number of classes.

7) Minimum Distance Classification and Clustering: Both the PL and FP are used in these operations. For 4 dimensions and 16 classes the processing time of the MD classification is 3.2s. By iterative application of MD classification, unsupervised or cluster classification can be obtained. The process takes approx. 10 frame times per class per cluster iteration which represents a speed improvement in excess of 100-fold over the host (IPC) implementation.

8) Fast Fourier Transform: The 2-D complex FFT is implemented using special microprogrammed instructions on the concurrently running IPC and FPP processors. The processing time is approximately 4 minutes for a 512x512 pixel image.

A TEXTURE ANALYSIS ARCHITECTURE USING ASSOCIATIVE MEMORIES

Helder J. Araújo [(1)], Anibal T. de Almeida[(1)]

Abstract

The concept of associative memory is used for the definition of a parallel architecture suited for texture analysis. Using associative memories, a representation of the images in an alternative space is built up. In this space, projections of the same class of textures ideally cluster. For recognition the unknown texture is projected into the alternative space and classified in the class whose cluster is closest to its projection.

1. INTRODUCTION

Some of the visual qualities used to distinguish textures are coarseness, directionality, linearity, roughness, and density [1]. Textures are usually classified into two types: structural and stochastic [2]. Structural textures can be considered as ordered arrays of identical subpatterns. Stochastic textures are textures where a subpattern or dominant repetition frequency are not recognizable. As a result texture analysis methods are also usually classified as structural and stochastic or statistical.

Statistical methods use statistical measures of the gray level spatial distribution to characterize and classify image textures. Many of the statistical approaches are based on the earlier results of Julesz [3]. According to these results our visual system is not sensible to differences in third and higher orders statistics of the gray level spatial distribution. Structural approaches usually must be "tuned" to the particular types of textures to be analysed [4]. Usually structural approaches can only be used when primitives and placement rules can be extracted [5]. A different approach, called spatial-statistical, was proposed by Laws [6]. This approach is based on the evaluation of the so-called "texture energy filters". A common feature to all the above mentioned approaches is their heavy computational cost.

We propose a new method of texture analysis and recognition which is inherently parallel and aimed at an hardware implementation, which has the advantage of enabling real time recognition (at frame rates).

2. THE BASIS OF THE APPROACH

Based on the concept of associative memory an alternative representation of the images can be built up. For this, a high-dimensional space is used. In this space images from the same class of textures are mapped into the same region. If the space's axis are appropriately chosen the projections of the images of the same class cluster. Recognition is performed by projecting the unknown texture image in this space. The unknown texture is classified in the class corresponding to the cluster nearest to the projection. A threshold can be defined to prevent misclassification when the minimum distance to the clusters is too large.

In our approach the projection of the images into the high-dimensional space is performed by using the contents of each image as memories' addresses. Each memory device corresponds to a particular axis of the high-dimensional space. The value of the address of a particular memory device, generated by a particular image, is the value of the projection of that image into the axis corresponding to that memory device. Based on these concepts several architectures can be defined.

(1)
Electrical Engineering Department, University of Coimbra, 3000 COIMBRA, PORTUGAL

The differences in these architectures result from the different possibilities of associating the contents of an image as memories' addresses.

2.1 PIXEL-BASED IMPLEMENTATION

The idea behind this approach is to have the image contents addressing memories. In this case (and for each texture class) we take each pixel of the image and use it to address a memory device. This memory device is one-bit wide. Learning consists in writing 1's in the memory positions corresponding to the values of the pixels (assuming that their are filled initially with 0's). There is one set of such devices for each one of the classes. For a 16 gray level image, each pixel addresses a memory device with 16 positions. For recognition the memory devices' outputs are associated in n-tuple samples and taken as inputs of logical gates. In our case these gates are AND's. The inputs of these gates can be the outputs of memory devices corresponding to linearly adjacent pixels or to randomly taken pixels. However the same image addressing scheme must be used during learning and recognition. The outputs of the AND gates are summed. During recognition there is one such structure for each class. Recognition is performed by addressing in parallel the network structures corresponding to each one of the classes. The unknown sample is classified into the class which had the greatest response. This corresponds to classifying the sample into the class closest to its projection in the alternative space. A threshold value for the responses can be defined so that a sample is not classified if the greatest response is smaller than the threshold. This implementation has a major drawback for the ranges of the space's axis values are non-variable. This structure was simulated and the performances were rather poor.

2.2 BIT-BASED IMPLEMENTATION

We are now going to describe another implementation based on the same concept. This implementation overcomes the limitation of fixed ranges of the values represented by the axis of the space. As a matter of fact this structure constitutes an extension to gray levels of the model originally proposed by Bledsoe et al. [7] and implemented in hardware by Aleksander et al [8].

Let us consider a rather simple example : a 3*3 binary image. Let us also group the pixels into three sets of three pixels each, on a column-by-column basis, and use each one of these groups to address a memory device with eight positions. We will now teach this network with three patterns (which are examples of a single class) - the patterns of Fig. 1. In this case " to teach " means to write a "1" into the three memory positions addressed by the image contents (assuming that the memories are initially filled up with 0's). Assume now that the output terminals of the memory devices are connected to the inputs of an AND gate.

The system will recognize a pattern if the output of the AND gate is "1". The recognition phase of the system is activated by feeding the network with the pattern to be recognized and by "reading" the memory devices, which will output a "0" or a "1" according to the image contents (and also according to the patterns used in the training phase). As we have an AND gate at the decision layer of the network, a pattern will be recognized if all the three memory devices output a "1".

If the network was trained with the patterns of Fig. 1, then it will recognize (in addition to the training set) the pattern of Fig.2.

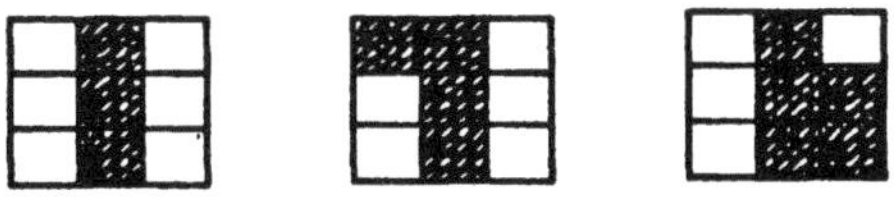

Fig.1

If we change the type of pixel grouping the recognition performances will be affected. If, for example, we group the pixels on a line-by-line basis (also into three sets of three pixels each), the set of patterns which will be recognized (for the same training set) will be different. The network's degree of generalization is affected by the mapping of the pixels into addresses of the memory devices.

To deal with the problem of multiple class recognition we can use one network similar to the one above described per class. The way according to which the n-tuples are constituted strongly affects the recognition performance of the system. For recognition the outputs of the memory devices are summed instead of being inputs to an AND gate. Again a sample is classified into the class corresponding to the greatest response.

A maximum response detector can be used to classify the unknown texture.

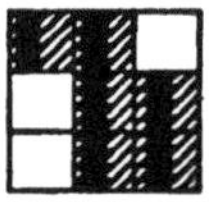

Fig. 2

In the case of texture images we must deal with the gray level information. To do so, the image is considered as a set of bits and the addresses of the memory devices are formed by taking n-tuples of bits. In this case the spacial meaning of the pixel association is lost. As a matter of fact in this type of implementation we pick up the bits to form the memory addresses without caring whether they belong or not to the same pixel. A major drawback of this approach is that all bits are considered contributing equally to the information stored in the discriminator. The contribution of each pixel's most significant bit should be different from the contribution of the pixel's least significant bit. Not doing so, the performances of this implementation are negatively affected as the least significant bits are more prone to noise and error than the most significant bits, and in considering all of them equally we are increasing the weight of the noise on the information provided to the discriminators.

2.3 PLAN-BASED IMPLEMENTATION

To deal with gray level images without the drawback of the above described approach, another architecture based on the concept of associative memory is possible.

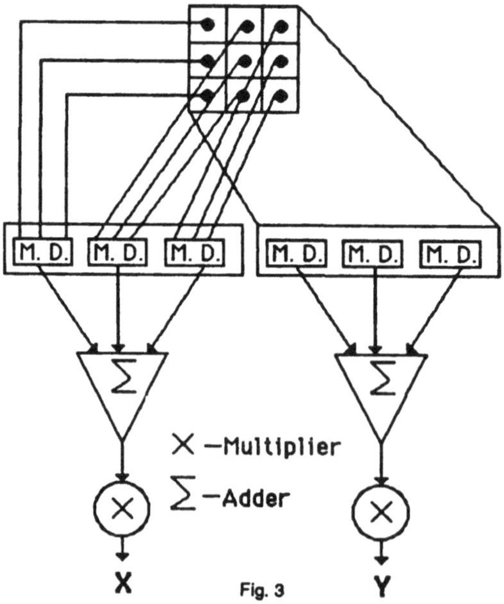

Fig. 3

In this case we consider the image made up of one-bit planes. For example if we have a 3*3 image with 16 gray levels, we can consider the image divided into 4 one bit images or planes. Now we make the projection of the image into the high-dimensional space plane by plane. At each one of these planes we have a structure similar to the one described in 2.2. At each one of them we have a discriminator per class. For recognition the outputs of the memories in each plane are added and multiplied by a factor proportional to the significance of the plane (Fig. 3- structure for one plane and two classes). Thus the contribution of each bit is proportional to its significance at the gray level dimension. The associations of bits to make up the addresses of the associative memories have now a spacial meaning as each bit belongs to a different pixel.

For each class the outputs of all levels are added, and next they are inputted to a maximum-response-detector. That is, the "vertical" association of the responses is made by class.

The three approaches above described have a common weakness: the memory devices only register the occurrence or not of the addresses. The information relative to the frequency of occurrence is lost. However, to deal with the frequencies of occurrence, the architecture would be more complex. Memory devices should have multiple-bit cells and the contents of each cell should be increased by one unity each time the corresponding address occurred. Much better results can be obtained when the frequency of occurrence of the addresses is registered. This corresponds to a weighting procedure in the alternative space. Even if the same address occurs in different textures, it is unlikely that its frequency of occurence is the same for all of them.

The results obtained with the model that keeps track of the frequency of occurrence of the addresses were significantly better than those obtained with the simpler model.

3. RESULTS

The three above described implementations have been simulated. This section describes a few of classification experiments done comparing the performances of the three approaches.

The data set used in this study consisted in five different classes of textures (A, B, C, D, E). These textures were samples of five different fabrics.

On the first experiment the three models were tested with a n-tuple of 10, using only the six most significant bits of each pixel. To form the addresses of the memories we used a random number generator. Numbers were generated in a total equal to the number of pixels (in the cases of the pixel-based and plane-based models) or in a total equal to the number of input lines of the discriminators (in the case of the bit-based model).

On this experiment we used a learning set of 128 samples per class. These samples were taken on only 32 different positions. The test set was constituted by 64 samples per class. The contingency table for the classifications, in the case of the pixel-based implementation is Table I.

Table I-Contingency table for the classification in the first experiment (pixel-based implementation).

CORRECT CLASS	CLASSIFIED AS A	B	C	D	E
A	52			12	
B		64			
C			45	19	
D				64	
E				39	25

These results can be considered poor as the different positions of the samples in the test set were the same as those of the learning samples.The results in the case of the bit-based implementation and plane-based implementation are displayed in Table II .

With the bit-based implemetation as well as with the plane based implementation all the test samples were correctly recognized. To compare the results of the bit-based model with the results of the plane-based model we considered the differences between the two largest responses of the discriminators. As a parameter we used the ratio between these differences and the theoretical maximum possible difference. For the comparison we considered two values: the maximum difference and the minimum difference registered with the 64 test samples. Thus the results could be compared with those of the plane-based implementation. These results are presented in Table II.

Table II-Maximum and minimum separations between the two most voted classes in the case of the bit-based implementation and plane-based implemetation.

	Bit-Based Model		Plane-Based Model	
	Maximum Diff.	Minimum Diff.	Maximum Diff.	Minimum Diff.
A	28%	1%	71%	24%
B	30%	17%	60%	36%
C	20%	6%	55%	19%
D	11%	4%	59%	25%
E	19%	11%	46%	12%

Comparing the results of the plane-based model with the corresponding results of the bit-based model we conclude that the best implementation is the plane-based one.

In another set of experiments we tested the behaviour of the plane-based model in the case where the positions of the test samples were not included in the learning set. In this experiment we used a n-tuple of 12, and considered the six most significant bits per pixel. The learning set was constituted by 384 images per class. The test set was constituted by 128 images per class. All the images were taken in distinct positions. In this case the contingency matrix was the one of Table III.

Table III- Contingency matrix for the plane-based implementation (free positions of the samples).

CORRECT CLASS	*CLASSIFIED AS* A	B	C	D	E
A	125		2	1	
B		128			
C		3	120	5	
D	1		61	65	1
E		14	76	8	30

These results are not satisfactory. Specially the results concerning the classification of class D and E samples are rather poor. To improve the recognition rates the images were preprocessed. Several filters and sequences of preprocessing operations were tested. The best classification results were obtained with the application of the following sequence of pre-processing operations to all samples:

1- Adaptive histogram equalization; 2- Low-pass filtering; 3- Edge detection .

The contingency matrix for the data set corresponding to the results of Table III, is presented in Table IV.

Table IV- Contingency matrix for the plane-based implementation (free positions of the preprocessed samples).

CORRECT CLASS	CLASSIFIED AS A	B	C	D	E
A	128				
B		128			
C			125	3	
D			5	123	
E			7		121

The worst classification rate (94%) concerns class E. The overall accuracy of classification of the test set was 97.7%.

4.CONCLUSIONS

We described a special architecture for texture analysis and recognition that is inherently parallel and aimed at an hardware implementation, which has the advantage of enabling real time recognition (at frame rates).This architecture proved to achieve high recognition rates of textures. Several implementations were considered and compared. One of them proved to be clearly superior to the others. Preprocessing of the images was used to achieve higher recognition rates.

REFERENCES

[1] H. Tamura, S. Mori, T. Yamawaki, "Textural Features Corresponding to Visual Perception", IEEE Trans. on Systems, Man, and Cybernetics, Vol. SMC-8, No. 6, 1978.

[2] R.M. Haralick, "Statistical and Structural Approaches to Texture", Proceedings of the IEEE, Vol. 67, No. 5, 1979.

[3] B. Julesz, "Visual Pattern Discrimination", IRE Trans. on Information Theory, Vol.8, No 2, February 1962.

[4] J.-G. Leu, W.G. Wee, "Detecting the Spatial Structure of Natural Textures Based on Shape Analysis", Computer Vision, Grap. and Image Processing, Vol. 31, pp. 67-88, 1985.

[5] L.V. Gool, P. Dewaele, A. Oosterlinck, "Texture Analysis Anno 1983", Computer Vision, Grap. and Image Processing, Vol. 29, pp. 336-357, 1985.

[6] K. I. Laws, "Textured Image Segmentation", University of Southern California, USCIPI Report 940, January 1980.

[7] W. W. Bledsoe, C. L. Bisson, "Improved Memory Matrices for the n-tupple Pattern Recognition Method", IRE Trans. on Elect. Comp., Vol. EC-11, pp. 414-415, June 1962.

[8] I. Aleksander, T. J. Stonham, "Guide to Pattern Recognition Using Random-Access Memories", Computers and Digital Techniques, Vol. 2, No. 1, February 1979.

VISUALIZATION IN IMAGE ANAYSIS

G. Sommer, K. Voss, W. Ortmann [1)]

1. Introduction

Interpretation of complex multidimensional data with the help of interactive computer systems got widespread applications in the last two decades. Computer graphics, image processing and image analysis are developed to handle pictorial data. They are distinguished by as well their tasks, data structures as hardware equipment. But indeed they are not so away as it seems for a first look.

In *image processing*, new images from images are computed, which in any sense are more close to the mental models of the user. In *image analysis* or *computer vision*, the interpretation of images is done by the computers themselves. The image data are transformed to data structures with internal representations of structures of interest. From these internal representations, verbal descriptions are to be derived and new images are to be synthesized. *Computer graphics* transforms internal representations of data structures into image structures [1].

This indicates to common aspects of computer graphics and image analysis, which in the last decade indicate a convergence to a common hardware base in raster scan systems [5].

Visualization as a user interface plays a central role and is of common interest for image processing, image analysis, and computer graphics [10]. Contrary to this, visualization got only minor attention in the past.

There will be given a survey of improvement of visualization by considering the requirements to outline image processing tasks and the architecture of image processing systems hardware as well as the perception principles of the human visual system (HVS).

---

1) Friedrich-Schiller-University, Department of Technology
GDR-6900 Jena, Ernst-Thaelmann-Ring 32

2. Visualization in the context of image analysis

Image analysis is roughly partitioned within a flow diagram into different not independent subtasks, which are to be realized on different data structures [1],[2].

STEP 1: *image processing* - images are transformed into images

For the purpose of visual image interpretation, the processing chain is abandoned at this step. For the purpose of computer vision this is a preprocessing step (maybe negligible), which results for instance in a pile of images with intrinsic properties in analogy to cortical transforms of the HVS [16],[7].

STEP 2: *model representation* - images are transformed into representations of scene models

These models represent the context of images, which means descriptions of structures and their relations [1]. This step includes segmentation of images into regions which get meaning for instance by feature extraction and classification, and the merging of that regions into aggregations. In the analysis of temporal or spatial sequences, it includes the reconstruction of interframe correspondences. To support the following step of model presentation, representation is e.g. symbolic with attributes characterizing topologic, geometric and other properties [14].

STEP 3: *model presentation* - internal representations of models are transformed to external presentations

These presentations are for instance verbal descriptions (e.g. medical diagnosis), pictorial visualization of 2D or 3D scene interpretations (e.g. to watch for changes in some sense) or reactions to start another process of action (e.g. robot vision). Visualization is the process to display images on an external medium (e.g. TV monitor or hardcopy) [6].

Thus, there are two reasons to visualize, which are related to the task of the analysis:

1. visualization to support human visual interpretation of images [15]
2. visualization to support interactively human decisions concerning machine interpretation of images [3].

Therefore, image processing software in general should support both a constrained coupling and a decoupling of processing and visualization. The way to do this concerns an appropriate architecture of image processing systems and an appropriate organization of image memory.

* Processing and visualization require each separate memories. An essential part of visualization is a mapping from any processing memory to visualization memory supported by hardware (special processor). In the case of lack such support, the visualization memory may be virtually and a part of the processing memory. Processing results in mapping between different parts of the processing memory.

* The visualization memory is organized to support vectorized (image) data. The processing memory is organized to support either vectorized or list data. In the case of prime and intrinsic images each pixel (x,y) of the processing memory possesses a numerical value g, $0 \leq g \leq GMAX$ which in general is called gray value. Pixels [x,y] of label images possess non-numerical values $l \in \{l_0, l_1, \ldots, l_{MAX}\}$, which can be mapped into the set $\{0,1,\ldots,MAX\}$. They express semantic aspects of image analysis and may be used to mask prime and intrinsic images.

* Pixels {x,y} of the visualization memory yield from a mapping of gray values g or labels l into intensity values to drive the display (or other device). The transformation rule of this mapping concerns dynamics and non linearity of the intensity scale or color scale and the scale of spatial resolution. Planar geometric transforms are of interest, too. Intensity values of labels are overlayed that of gray values by higher priority. Any update of the visualization memory has to consider the separate update from gray values and labels.

* The visualization of 3D scenes requires some additional processing steps in the mapping process [15]. These rendering steps include geometric transforms from an object centered to an user centered coordinate system, hidden surface removal and surface shading.

* The structure of list data is related to the tasks of symbolic processing or computer graphics. In the context of visualization the question is open whether we need standards as in computer graphics.

Mapping of image data for processing and visualization are effectively supported by a proper organization of image memories on the user level. If there is an image processing memory of size MxN it can be organized into a gray value part $M_g x M_g$ and a label part $M_l x M_l$. Each of them is organized into a matrix of totally Z matrices, each of size XxY. Any pixel (x,y) or [x,y] is restricted to the range XxY. Each matrix gets its own number z_g or z_l. So the effective pixel address is $\{x,y;z_g\}$ or $[x,y;z_l]$. Processing within one selected matrix of number $z_g=z_g^{akt}$ or $z_l=z_l^{akt}$ needs only coordinates (x,y) or [x,y]. It is worth, giving the same organization to the visualization memory of dimension $M_v x M_v$, which results in the pixel address $\{x,y;z_v\}$.

This architecture gives the possibility of using not only special processors for image processing but also for rendering internal representations of pictorial data. They have to lower the high computational loads not only in 3D visualization but also in the adaption of displayed images to the HVS.

The architecture also permitts more logical separation of image analysis software from those of (partially hardware related) image visualization.

3. Visualization in the context of perception

Visualized images play the role of user interfaces [9]. There are fundamental problems to optimize this interface. Generally we must state, that the task of adaptation of image presentation to the perception of the HVS is an unsolved problem [3],[10]. This may not wonder, because it requires to know the principles of visual perception on the whole. If we would know that process, we would also do better image analysis.

Generally, perception is determined by physical properties of displayed images and by the semantics of mental world models of the user. Therefore, visualization is a problem of psychophysics [10] and, in a strong sense, it has to be realized in the context of questions in mind for each image class separately. As invariancesin a more loose sense we can state, that the mental world model of the user is a world of continous coordinates and intensities respectively. As approximation it seems reasonably to project the whole problem to the subproblems of quantization of the display scale and intensity dynamics range.

** Spatial sampling

Because of the minimal distance of two points to be viewed as separate points is in the order of 1 minute per arc, the pixel size of the display is determined by this value. In the context of interactive image processing (32 cm display diameter, 50 cm viewing distance) this results in 2000 pixel in each dimension of the display, which wrecks today from technical reasons. But also reasonable display dimensions of 1000 or 500 pixels often require resampling of measured and stored images. To do this by zooming alone (which is the common method), results in artefacts due to the perceived edges of pixel blocks. Instead of this, resampling must be coupled with interpolation [8]. The best known interpolation methods with minimal sideband energy in frequency space are cardinal splines.

B-splines are in that sense suboptimally but are very simple to realize.

In this context, also methods of antialiasing which are developed in computer graphics have to be seen [12]. They result in impressions of computer graphic presentations, which favour continuity.

** Intensity scaling

The mapping of gray values onto display intensities ought to be a transformation into perceived intensities. Visualization concerning intensity scaling breaks into two steps: contrast enhancement and linearization of perceived intensities.

Contrast enhancement has to minimize the transfer losses in the display of recorded intensities [4]. A well known method is histogram equalization. Because of intensity differences should increase with the importance of seeing the differences, contrast enhancement should be done local adaptively [17]. The output of this mapping is the so called scale indicator intensity [10].

Linearization of perceived intensities with respect to the scale indicator intensity results in the independence of the perceived contrast of the display device. To realize this, the relation between perceived and physical driving intensity of the display must be known. This can be measured by ROC experiments [13] concerning that change in driving intensity that results in just perceivable different displayed intensities [11]. Knowledge of this relation permits to transform the scale indicator intensity to physical driving intensity, so that the relation between scale indicator intensity and perceived intensity becomes linearly.

To look away from theoretical and methodical difficulties, visualization in the context of perception needs an architecture of image processing systems, which separates strongly visualization from processing of images. The mapping of recorded to displayed intensities requires separate processor support.

4. Literature

[1] Ballard, D.H., Ch.M. Brown: Computer Vision.
Prentice Hall,Inc., Englewood Cliffs, New Jersey, 1982

[2] Barrow, H.G., J.M. Tenenbaum: Recovering intrinsic scene characteristics from images.
in: A.R. Hanson, E.M. Riseman (eds.):Computer Vision Systems, Academic Press, New York, 1978

[3] Caelli, T: Digital image-processing techniques for the display of images and modeling of visual perception.
Behav. Res. Meth., Instr. & Comp. 18 (1986) 6, 493-506

[4] Cormack, J., B.F. Hutton: Minimisation of data transfer losses in the display of digitised scintigraphic images.
Phys. Med. Biol. 25 (1980) 2, 271-282

[5] Jackel, D.: Raster-Scan-Displays für Computer-Graphik und Bildverarbeitung.
Handbuch d. Modern. Datenverarbeitung 127 (1986) 23-42

[6] Jackel, D.: Eine Rechnerarchitektur zur Visualisierung und Verarbeitung von 'cellular space'-Repräsentationen.
Dissertation, Techn. Univ. Berlin, 1987

[7] Marr, D., E. Hildreth: Theory of edge detection.
Proc. Royal Soc. London B 207 (1980) 187-217

[8] Parker, J.A., R.V. Kenyon, D.E. Troxel: Comparison of interpolation methods for image resampling.
IEEE Trans. Med. Imaging MI-2 (1983) 1, 31-39

[9] Pentland, A.P.: Toward an ideal 3-D CAD system.
Proc.SPIE 758, Image Understand. and the Man-Machine Interface (1978) 165-171

[10] Pizer, S.M.: Psychovisual issues in the display of medical images.
in: K.H. Höhne (Ed.): Pictorial Information Systems in Medicine.
Proc. NATO ASI (1985), Springer-Verl., Berlin, Heidelberg, New York, 1986, 211-234

[11] Pizer, S.M. F.H. Chan: Evaluation of the number of discernible levels produced by a display.
in: R. Di Paola, E. Kahn (eds.): Information Processing in Medical Imaging, Edition INSERM, Paris, 1980, 561-580

[12] Purgathofer, W.: Realistische Bilder.
Handbuch der Modernen Datenverarbeitung 127 (1986) 43-61

[13] Rajalakshmi, P.K., A.V. Lakshmanan: A comparison of the methods for assessing the performance of the display system of a clinical radioisotope imaging device.
Phys. Med. Biol. 30 (1985) 4, 293-301

[14] Shapiro, L.G., R.M. Haralick: Relational matching.
Applied Optics 26 (1987) 10, 1845-1851

[15] Tönnies, K.D.: 3D-Representation der Morphologie von anatomischen Objekten durch Approximation ihrer Objekte.
Dissertat., Techn. Univ. Berlin, 1987

[16] Watson, A.B.: The cortex transform: rapid computation of simulated neural images.
Comp. Vis., Graph., Image Processing 39 (1987) 3, 311-327

[17] Zimmermann, J.B.: The effectiveness of adaptive contrast enhancement
TR 85-029, Dept. Comp. Sci., Univ. North Carolina, Chapel Hill, 1985

AN EXPERT SYSTEM FOR THE APPLICATION OF IMAGE PROCESSING SUBROUTINES

Felix Grimm[1], Horst Bunke[1]

ABSTRACT

This paper presents an expert system which supports unexperienced users in the application of the SPIDER image processing software package. The system has been completely implemented and tested on a number of image processing problems.

Keywords: image processing, SPIDER package, expert systems, application of subroutines by means of expert systems.

1.0 INTRODUCTION

Image processing becomes more and more important in many disciplines of science and engineering. Much research work has been done and is actually done in this field. A large number of image processing algorithms have been developed in recent years. Some subroutine packages are available now containing a variety of such algorithms. An example is SPIDER (see [1]) which is a collection of more than 400 subroutines (including a large number of lower level routines). These routines have been written in Fortran and collected at the Electrotechnical Laboratory (ETL) in Japan.

In this paper we present an expert system for the application of the SPIDER package. This system is mainly intended for users without deep knowledge and experience in the field of image processing in general, and with SPIDER in paritcular. Some systems similar to ours have been proposed in the literature recently (see [2] - [7]).

2.0 SYSTEM OVERVIEW

The expert system SPICE (Subroutine Package for Image analysis Consultant Expert) described in this paper consists of a system kernel and a number of peripheral components. The kernel is divided into the knowledge base and the inference engine while the peripheral components include three interfaces, one to the user, one to the knowledge engineer, and one to the SPIDER subroutine package. The main emphasis of this paper will be on the system kernel. For a detailed description of the system, see [8].

Given an image analysis problem, such as finding the contours or the area of objects in an image, our system first asks the user some questions to find out his intentions (i.e. it tries to classify the problem). Then it selects, in an interactive dialog, a sequence of SPIDER routines, applies them to the input picture and displays the results after each step of processing. The user can accept or reject the obtained results. In the latter case, the system explores alternative ways of processing. Finally the sequence of SPIDER routines actually executed is written onto a file. From this file, it is possible to generate a program which can be used, from then on, for similar problems.

3.0 KNOWLEDGE BASE AND INFERENCE

The knowledge base of SPICE logically consists of three components containing knowledge about SPIDER subroutines including their parameters, knowledge about image analysis, and knowledge about general principles for calling and executing subroutines. This knowledge is represented by means

---

[1] Institut für Informatik, Universität Bern, Länggassstrasse 51, CH-3012 Bern, Switzerland

of rules and frames. Inference is based on forward and backward chaining of rules in combination with property inheritance of frames.

Our system contains knowledge about image processing in about the same extent as the SPIDER package (except the fields of orthogonal transforms and texture analysis). This knowledge describes problem solving strategies of image processing which include problem classification, decomposition of problems into less complicated subproblems which can be solved separately, selection of appropriate algorithms as well as fine tuning by assigning parameter values. An example of a rule used for **problem classification** is

IF the goal of processing is the extraction of numerical features
AND the kind of numerical features is location of lines
THEN the goal of the consultation is location of lines.

The problem solution is subdivided into **subproblems** (subgoals) by means of rules like

IF the goal of the consultation is location of lines
AND the user is interested in boudary lines based on region segmentation
AND region segmentation has not yet been performed
THEN the actual subgoal is region segmentation.

Similar rules deal with the subdivision of subgoals into steps. Given the purpose of the actual step, an **algorithm** has to be selected next. This is done by means of rules like

IF the purpose of the actual step is region segmentation in a binary picture
THEN the actual algorithm is extraction of connected components.

The knowledge about SPIDER relates algorithms of image processing with corresponding names of SPIDER subroutines. For those cases where there is only one subroutine for an algorithm, a table is used. Otherwise further constraints are taken into account, using rules like the following one:

IF the actual algorithm is affine transform
AND linear interpolation is to be used
THEN the actual subroutine is AFIN1.

The SPIDER routines and their parameters are described by means of frames while rules are used for finding parameter values.

Our system is not restricted to sequential execution but also allows **iterative flow of control** and includes a **backtrack mechanism**. Subgoals and steps may be repeated, if necessary. For example, if the system has performed an image enhancement like smoothing and the user is still not satisfied by the obtained image quality, another image enhancement will be scheduled. This is achieved by the following two rules:

IF the previous subgoal has been image enhancement
AND the user is still not satisfied by the obtained image quality
THEN the previous subgoal is to be repeated.

IF the previous subgoal is to be repeated
THEN the actual subgoal is the same one as the previous subgoal.

In the above situation, the same subgoal is applied once more on the picture actually generated (iterative flow of control).

In other cases the user may reject an obtained result. In such cases, the system has to explore alternative ways of processing starting from a former picture (i.e. it has to perform a backtrack). A region segmentation process, for example, may split up objects into a large number of meaningless regions as a consequence of insufficient image smoothing. If the user considers the segmented picture as too much splitted, the system undoes the region segmentation, performs image smoothing on the original picture and starts region segmentation again, this time on the smoothed picture. The backtrack mechanism for this situation is described by rules like the following ones:

IF there are too many meaningless regions in the segmented picture
THEN the system has to perform a backtrack
AND the reason of the backtrack is insufficient image smoothing
AND all conclusions in the dynamic knowledge base starting from image enhancement are to be retracted.

IF the actual subgoal is image enhancement
AND a backtrack has been performed
AND the reason of the backtrack is insufficient image smoothing
THEN the kind of image enhancement to be performed is image smoothing.

More details about the knowledge base of SPICE can be found in [8] and [9].

4.0 IMPLEMENTATION

As already mentioned in the previous sections, our expert system SPICE uses rules and frames for knowledge representation. Besides, there are two tables (for relating image analysis algorithms with SPIDER subroutines). Inference is mainly controlled by backward chaining, but there exist also some forward chaining rules. The expert system shell TWAICE by Nixdorf has been used for the implementation. It directly supports tables and rules under backward and forward chaining inference. As frames are not supported by TWAICE, frame-based inference including an inheritance mechanism has been implemented in Prolog.

The actual implementation of our system SPICE covers about 90 subroutines, which is approximately one third of the higher level SPIDER routines. It contains about 500 rules and about 200 frames. There are also some Prolog and C procedures which are triggered by rules or events. These programs are used for property inheritance of frames, interface to the subroutines, determination of subroutine parameters, file handling, backtrack mechanism, etc. The interface program allows subroutines to be executed under the control of the expert system and parameters to be passed from the expert system to the subroutines and vice versa. The current prototype of our system has the following limitations:

- Only a part of the field of image analysis and of the SPIDER package are implemented in SPICE.
- An user of our system must have some basic knowledge about image analysis, as the system often asks questions about the next task to be done. However, SPICE provides also some help facilities like explanation of questions asked by the system which is displayed upon the user's request.
- The graphics interface of the current version of SPICE is very limited. It only allows pictures to be displayed, but no direct interactions on the pictures (like specification of coordinates).

5.0 EXPERIMENTAL EXAMPLE

This example shows the processing of a picture containing 7 coins (Fig. 1, Picture 1) under SPICE. The following parts of a dialog between an user and the system are original up to translation from German to English. After some initialization work, the user is asked to specify the processing goal:

```
(5) What is the goal of processing this picture?
    1: another processed picture
    2: extraction of numerical features
    3: approximation of image objects
    4: symbolic representation of image objects
    5: object recognition
> 2
```

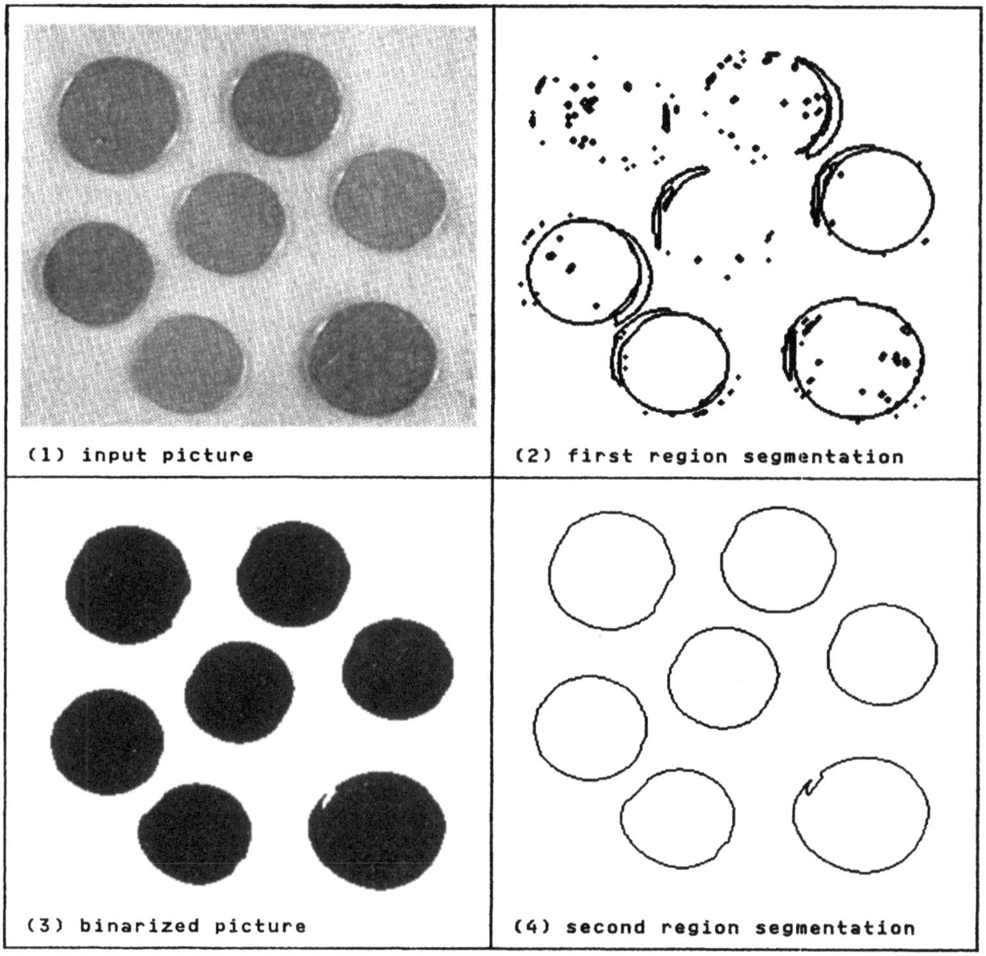

Figure 1. Pictures of the example

```
(6) What kind of numerical features?
     1:  location of lines
     2:  position of regions
     3:  coordinates of shrunk components
     4:  features of regions
     5:  location of elementary patterns
     6:  statistics
     7:  other features
> 2; 4

(7) What kind of feature of regions are you interested in?
     1:  area
     2:  perimeter
     3:  size
     4:  compactness
     5:  moment
     6:  distances from center of gravity to boundary points
     7:  slope and curvature of boundaries
> 1
```

coin	x-coordinates	y-coordinates	area (number of pixels)
coin1	55	166	3634
coin2	67	68	4472
coin3	111	223	3409
coin4	138	135	3418
coin5	172	61	3685
coin6	226	220	4746
coin7	237	120	3438

Figure 2. Features of regions computed by SPICE

(8) How shall the regions be described?
    1: starting point of the border
    2: circumscribing rectangle
    3: center of gravity
> 3

With this information the system has identified two goals: position of regions (described by the center of gravity) and features of regions (area). The system now asks the user if image enhancement or image transformation is necessary. As our answer has been "no", the system directly starts region segmentation by means of the SPIDER routine RCRS. After region segmentation and boundary detection, Picture 2 which contains 315 regions is displayed. The system now asks:

(13) Do you think the picture is too much splitted up?
    1: yes
    2: no
> 1

Because of this answer, the system performs a backtrack and restarts processing with the input picture (Picture 1). It automatically selects image enhancement as the fist subgoal and executes the subroutine MEDI for image smoothing by median filter. As we have accepted the system's proposal to binarize the picture, it uses suitable routines for automatic threshold selection and binarization (Picture 3). In the following subgoal, region segmentation is done by means of the subroutine CLAB (Picture 4) yielding 7 regions. Finally the coordinates of the centers of gravity as well as the areas of the regions are computed. They are shown in Fig. 2 (the top left most point of the picture is the origin of the coordinate system; the size of the picture is 270 rows x 300 columns). In the present version of the system it isn't possible to overlay the coordinates given in Fig. 2 directly on the image. However, such a facility could be added without any major problem. The list of the successfully applied subroutines is MEDI, HIST1, THDS2, SLTH1, CLAB, BFLP1, CGRV1, AREA1.

More examples of pictures processed under SPICE can be found in [8] and [9].

6.0 CONCLUSIONS

An expert system for the application of the SPIDER image processing software package has been presented in this paper. Our system is beneficial to users without any knowledge of the SPIDER package and without great experience in image processing. The discipline of image analysis is fast growing and expanding into many new application areas like robot vision for industrial assembly and quality control, pictorial input of printed documents to computers, medical diagnosis, and others. So systems like ours are of great practical interest. There is no doubt that an expert system like SPICE can enhance the application of standard library programs with respect to ease of use, efficiency, and economy.

As already mentioned in section 3, the knowledge base of SPICE contains three different kinds of knowledge. Accordingly, we have logically divided our knowledge base into three parts. In this way a great degree of flexibility and versatility is achieved. For example, if the system is to be adopted to another image processing subroutine package, only the part describing SPIDER has to be changed. Furthermore, that part of the knowledge which deals with general strategies of problem solving by means of subroutine packages has been implemented as an independent module and can therefore be used in conjunction with other problem areas and other subroutine packages. Together with the inference engine and the interface to the subroutine package, it can be considered as a special shell for building expert systems for the application of subroutine packages. Regarding this feature we feel that our system is based on a more general approach than other similar expert systems for image processing which have been reported in the literature recently ([2] - [7]). Since more and more non-specialists become increasingly attracted by standard software packages in various domains, shells of the above type potentially have a broad spectrum of applications.

7.0 ACKNOWLEDGMENT

We wish to thank Nixdorf Computer AG Switzerland for supporting our project by making us available a Nixdorf Targon /31 computer system including the expert system shell TWAICE.

8.0 REFERENCES

[1] H. Tamura, S. Sakane, F. Tomita, N. Yokoya, M. Kaneko, K. Sakaue: Design and Implementation of SPIDER - a Transportable Image Processing Software Package, Computer Vision, Graphics, and Image Processing 23, 1983, p. 273 - 294

[2] T. Matsuyama: Expert Systems for Image Processing: An Overview, in I.T. Young et al. (ed): SIGNAL PROCESSING III: Theories and Applications, Elsevier Science Publishers B.V. (North-Holland), 1986, p. 869 - 872

[3] K. Pfitzner, H. Strecker: XRAY - An Experimental Configuration Expert System for Automatic X-ray Inspection, 9. DAGM-Symposium: Mustererkennung 1987, Informatik-Fachberichte 149, Springer Verlag, p. 315 - 319

[4] J. Hasegawa, H. Kubota, J. Toriwaki: Automated Construction of Image Processing Procedures By Sample-Figure Presentation, Proceedings of the 8th Int. Conference on Pattern Recognition, Paris, October 1986, p. 586 - 588

[5] K. Sakaue, H. Tamura: Automatic Generation of Image Processing Programs by Knowledge-Based Verification, IEEE Proceedings on Computer Vision and Pattern Recognition, San Francisco, California, 1985, p. 189 - 192

[6] H. Tamura, K. Sakaue: DIA (Digital Image Analysis) - Expert System: An Approach to Future Vision System Design, Presentation at Int. Symposium on Image Processing and its Application, Tokyo, Jan. 18 -21, 1984

[7] T. Toriu, H. Iwase, M. Yoshida: An Expert System for Image Processing, Fujitsu Scientific and Technical Journal 23/2 (1987), p. 111 - 118

[8] F. Grimm: Expert Systems for the Application of Subroutine Packages and an Example of an Expert System for Image Analysis (in German), PhD thesis, University of Berne, 1989

[9] F. Grimm, H. Bunke: An Expert System for Software Configuration and its Application to Computer Vision, Proceedings of the SPIE Conference on Applications of Artificial Intelligence VII, Orlando, Florida, USA, 27 - 31 March 1989

IMAGE DATABASE MANAGEMENT SYSTEM. APPLIED THEORIES, TOOLS AND DECISIONS

P. Stanchev[1] – F.Rabitti [2]

In this paper we address the problem of building a database system for images (IM_DBMS), in which the images can be accessed by a partial description of the image content. The approach allows a limited automatic analysis, using ARG techniques and a rule-based system for images belonging to a domain described in advance. The semantic objects, recognized in the images during the analysis process, are interpreted according to the Theory of evidence. The image query processing is based on special access structures generated from the image analysis process.

1. INTRODUCTION

When large amounts of images have to be managed in a computer system, the need to apply the database technology naturally arises. In the last decade, much of the work in the field of "image databases" appeared in the proceedings of the "IEEE Workshops on Pictorial Data Description and Management", starting from 1977, and in some other series such as "Computer Graphics and Image Processing", "Computer Vision, Graphics, and Image Processing", "Image and Vision Computing", etc. Some essential sources are: the work edited by Chang and Kunii [7] and the collections of papers on pictorial applications and information systems in [5], [6], [10], [12], etc. The valuable survey of Tamura and Yokoya [20] includes insights into many actual approaches, as well as descriptions of several systems, such as the Graphics-oriented Relational Algebraic Interpreter (GRAIN), the Relational Database system for Images (REDI), the Database system of Microscopic Cell Images (IDB), etc. Another work [8] presents a survey of seven commercial systems, currently available and their software capabilities: Xerox's 8010 Information System, TERA's Automated Records Management Systems, Teknotron's Systems, Scitex's Response 250, Toshiba's Document Image Filing System DF2100, CCA's Spatial Data Management System, IIS's System 600 series of software products.

But, most of the systems, appearing under the heading of image databases are often image systems without full database functionalities or database systems not directly dealing with images [13]. In fact, most of the existing systems are application specific, that is, the way in which images are stored, organized and retrieved is specific of a certain application and cannot be generalized to different applications.

The main conceptual problem in dealing with images derives from the difficulty to exactly define and interpret the content of images. Images can be very rich in semantics, but are subjected to different interpretations according to the human perception of the application domain. On the one hand, it is difficult to recognize the objects (with the associated interpretation) contained in an image, on the other hand is difficult to determine and represent the mutual relationships among these objects, since they form structures which vary greatly from image to image.

For the problem of image retrieval by content, one could think to apply Data Base Management System (DBMS) or Information Retrieval System (IRS) techniques. However, with respect to DBMS's, it is difficult to recognize regular structures of objects contained in images, and then organize image instances into a limited number of types, to which the interpretation is associated. This is the approach required by the strictly typed data models adopted in database systems [21]. In IRS, instead, a free formatting of text is allowed, usually respecting some loose hierarchical structuring in sections, subsections, paragraphs and sentences. These systems do not attempt to understand the text (unless some expert system approach is adopted), but still allow an effective retrieval on text. In fact, as opposed to image objects, they can exactly recognize words (as ASCII patterns), on which they base their retrieval capabilities with the possible help of a thesaurus to support synonyms [18]. This is

[1] Institute of Mathematics, Bulgarian Academy of Sciences, G. Bonchev Str., 8 - 1113 Sofia (Bulgaria)

[2] Istituto di Elaborazione della Informazione, National Research Council, Via S. Maria, 46 - 56100 Pisa (Italy)

possible, in case of text, because a common semantic is associated to the words used in the natural language. Hence, both DBMS and IRS approaches cannot be directly applied to image retrieval.

In addressing the problem of image retrieval of stored images, if we want to think of a system trying to do for images what DBMS and IRS do for formatted data and text, we must accept some indeterminantness, characteristic of images, and then deal with the inaccuracy introduced by this fact. In [14] an approach based on Fuzzy set theory has been applied to the analysis and description of pictorial images. Certainty factors for the recognition of objects inside the images are computed using fuzzy logic rules.

A major practical problem for pictorial images is element recognition. The first step in the image analysis and recognition process is the decomposition of the images into relevant and identifiable elements, which are the basic components which will constitute the building blocks of the image structure.

In this step, often called segmentation in pictorial image processing [1], the image space is partitioned into meaningful regions, corresponding to image elements. After the segmentation, the system must recognize the tentative elements in the image, matching them with the pictorial representation of the elements to be searched. In this process, different variations as changes in size, rotation, translation (eg. using discrete Fourier transforms), can be attempted. An additional problem arises when partial element overlapping occurs in the image.

The result of this process should be the set of basic elements recognized and their relative positions. However, this is a highly computing intensive process which often requires special hardware, such as array processors, exploiting the inherent parallelism of the algorithms in order to have acceptable response times. In the end, the system might even not be able to exactly identify the single elements.

In this paper, a database system for images, IM_DBMS, is presented. It supports the analysis and retrieval of images. It is required that the images belong to a specific domain which must be described in advance to the system. A limited automatic analysis of the images is performed before storing the images in the database. This process is accomplished using a rule-based system. The interpretation of the content of the images is based on the Dempster-Shafer theory of evidence [11].

2. THE IM_DBMS SYSTEM

The main data "objects" that are input, analysed, stored, retrieved in IM_DBMS are bit-map images, obtained through image scanning process or graphical images, obtained by the tools of a graphical editor.

Figure 1. shows the architecture for the IM_DBMS. In the system we can identify six inter-related phases: domain description, image input and editing, image analysis, image storage, image retrieval, and image output and communication.

2.1. Domain Description in IM_DBMS

The purpose of this phase is to describe the characteristics of the application domain of the images to be stored and retrieved. The domain description function supplies with initial information the various phases of the image creation and analysis. It comprises:

a) establishment of the domain basic elements, for the puproses of image elements understanding. Domain basic elements are the basic components which will constitute the building blocks of the image structures. Image elements are domain basic elements established during the segmentation and recognized during image element understanding phase.

b) describing the possible relations and attributes of the basic elements, for the purposes of object recognition. Objects are meaningful for the application domain basic elements constructions;

c) indication of representative images, for the purposes of the image interpretation and clustering.

2.2. Image Input and Editing

The puprose of this phase is to enter the image into the computer storage. Bit-map images are

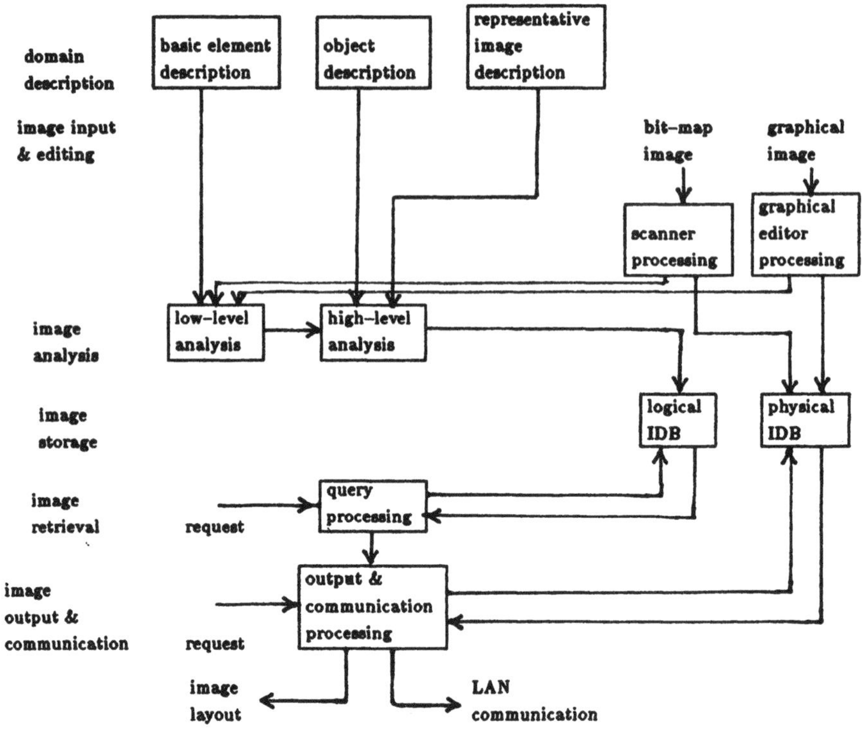

Figure 1. *Architecture for IM_DBMS*

acquired by a scanner and are transformed in image binary array form. Graphical images are created by a graphical editor and are transformed in a set of graphical editor primitives.

2.3. Image Analysis

The process of image analysis is concerned with the global analysis of the images. It could be considered separated into two phases.

The first phase, usually called low-level image analysis, is concerned with the extraction of an adequate form of knowledge from the image data and representing it in some appropriate formalism.

In the second phase, called high-level image analysis, the actual, in the sense of image meaning comprehension, analysis of the image is performed, using the extracted representation. The global image representation is obtained.

2.3.1. Low-Level Image Analysis

The low-level image analysis accepts an image (bit-map or graphical) and recognizes the basic elements in it, their relative positions and the associated distinguishing attributes.

The low-level image analysis is constituted by the following steps:

a) image elements understanding

Since the number of the basic elements (eg. polilines, curves, etc.) can be very large in a single image (in the order of thousands) a very efficient approach is required for recognizing the basic elements which are meaningful in the application domain. They constitute the basic symbols which compose the semantic objects in the image. In this phase, it is not possible to adopt a rule system, based on a generalized inference mechanism with back-tracking, because of its computational complexity. We need instead more efficient and specialized algorithms (with polynomial computation complexity) even if we have to pay this with a description system less rich in semantic content. For this reason we have adopted an approach based on the Attributed Relational Graphs [9].

The ARG graph is a relational structure which consists of a set of nodes and a set of branches representing the relations between the nodes, as both nodes and branches may have some attributes assigned to them.

During the Image Element Analysis, an ARG representation of the image (in terms of basic elements of the application domain, thier relationships and attributes) is obtained and stored in the image description file.

2.3.2. High-Level Image Analysis

The high-level image analysis uses the information about the elements, contained in the image and their relative position, obtained in the low-level image analysis. It is constituted by the following steps:

a) object recognition

The result of this phase is a set of objects recognized in the image.

The purpose of this phase is to recognize more semantically meaningful objects from the basic elements derived and organized in the previous phase. This task is accomplished by recursively applying the production rules defined for the chosen application domain. An inference mechanism based on backward chaining tries to derive from the basic elements more general objects and to give a recognition degree to the object recognized. In this phase a generalized inference mechanism is used. Its computational complexity is acceptable at this point, since fewer objects (in the order of hundreds) are present in the image.

The inference process starts from production rules obtained from the ARG image representation. After this step, a sequence in the form (1) is obtained:

(1)
$$\{O_{1\ 1}(\mu_{1\ 1}, l_{1\ 1}), \ldots, O_{1\ s_1}(\mu_{1\ s_1}, l_{1\ s_1})\}, \ldots,$$
$$\{O_{n\ 1}(\mu_{n\ 1}, l_{n\ 1}), \ldots, O_{n\ s_n}(\mu_{n\ s_n}, l_{n\ s_n})\}$$

Such a sequence describes an image with n distinct physical objects. The unit $O_{i\ j}(\mu_{i\ j}, l_{i\ j})$ is a semantical representation of the physical object $i (i = 1, 2, \ldots, n)$ in the image in the j–th ($j = 1, 2, \ldots, s_j$) recognition (i.e. a semantic object). $\mu_{i\ j}$ and $l_{i\ j}$ are respectively, the recognition degree (RD) and the list of attributes of the i–th physical object in the j–th recognition.

b) image interpretation.

The image interpretation is constituted by a sequence of all interpretations of the objects found in the image.

Using a procedure similar to Barnett's scheme [2], based on the Dempster–Shafer theory of evidence [11], and fully described in [15], we convert the results obtained from the previous phase into a list of new structures, containing information for each object:

(2)
$$\Big\{O_{1\ 1}([Bel(O_{1\ 1}), 1 - Bel(\overline{O}_{1\ 1})], l_{1\ 1}), \ldots,$$
$$O_{1\ q_1}([Bel(O_{1\ q_1}), 1 - Bel(\overline{O}_{1\ q_1})], l_{1\ q_1})\Big\}, \ldots,$$
$$\Big\{O_{n\ 1}([Bel(O_{n\ 1}), 1 - Bel(\overline{O}_{n\ 1})], l_{n\ 1}), \ldots,$$
$$O_{n\ q_n}([Bel(O_{n\ q_n}), 1 - Bel(\overline{O}_{n\ q_n})], l_{n\ q_n})\Big\}$$

Here $q_i \leq s_i (i = 1, 2, \ldots, n)$. The function $Bel(O_{i\,j})$ $(i = 1, 2, \ldots, n,\ j = 1, 2, \ldots, q_i)$ is a belief function.

The belief function $Bel(O_{i\,j})$ gives the total amount of belief committed to the object $O_{i\,j}$ after all evidence bearing on $O_{i\,j}$ has been pooled. The function Bel provides additional information about $O_{i\,j}$, namely $Bel(\overline{O}_{i\,j})$, the extent to which the evidence supports the negation of $O_{i\,j}$, i.e. $\overline{O}_{i\,j}$. The quantity $1 - Bel(\overline{O}_{i\,j})$ expresses the plausibility of $O_{i\,j}$, i.e., the extent to which the evidence allows one to fail to doubt $O_{i\,j}$. The interval $[Bel(O_{i\,j}), 1 - Bel(\overline{O}_{i\,j})]$ is called belief interval.

In the expression (2), object interpretations with "low" belief (e.g., in the sense of interval mean value less than a chosen one) could be omitted.

c) **image clustering**

The image clustering process is, in principle, similar to the document, clustering of text documents used in Information Retrieval Systems [18]. The most significant classes of images in the application domain are defined in terms of representative images, one for each class. The image interpretations are clustered by comparing them with the class representative images. After this computation the clustering description of the image is expressed as a sequence:

(3) $$\mu_1, \mu_2, \ldots, \mu_p,$$

where μ_i is the membership degree of the image to the i-th class. .

2.4. Image Storage

The information obtained in the image analysis phase can be used to generate access structures on image content which can be used for efficient image retrieval. Access structures are mainly indices on the objects, contained in all the images, with the associated attributes, and clusters about the interpretations of the images. The images are stored in a physical image data base (physical IDB) and their descriptions in a logical image data base (logical IDB).

The image representation is stored in a file containing, either the graphical image, as sequence of the graphical primitives applied for its composition by the graphical editor, or the binary array representation, in the case of bit-map images obtaining.

The derived image information, resulting from the analysis phase (expressed in terms of the probabilistic model, as composition of objects at different level of complexity, with the associated interval of belief) is stored in an "image header", associated to the image file.

Access structures (that is, the image indices) are built for a fast access to image headers. Two type of indices are constructed:

* Object index. For each object a list OBJ_IND is maintained constituted by a: DOC_ID, OBJ_NAME, OBJ_NUM, ATT, VAL, OBJ_CEN, OBJ_CON, OBJ_OVER, BI1, BI2 where: DOC_ID is a pointer to the image identifier; OBJ_NAME is the object name; OBJ_NUM is the number of the object in the image; ATT and VAL are two arrays, with attribute names and attribute values; OBJ_CEN contains the coordinates of the object center; OBJ_CON is an array with the numbers of the contained objects; OBJ_OVER is an array with the numbers of the overlapping objects; BI1 and BI2 are the bounds of the associated belief interval.
* Cluster index. For each class, defined in the application, a list CLU_IND of elements (DOC_ID, CLU_NAME, DEGREE) is maintained, where DOC_ID is a pointer to an image identifier, corresponding to an image with a non-null degree of membership to this cluster, CLU_NAME is the cluster name, DEGREE is the value of the membership degree.

2.5. Image Retrieval

The developed in IM_DBMS language is a newly developed language for image retrieval by content.

According to the query language, the user specifies the image condition in the form: RETRIEVE IMAGES <image_clause>.

The <image_clause> contains a <cluster_clause> and/or a <object_clause>.

The <cluster_clause> is an "OR" combination of <cluster_predicate>s, each of the form: <cluster_name> <cluster_degree> . This predicate indicates that the images in the database with a similarity to the named class higher than the <cluster_degree> should be retrieved (as requested in the "OR" expression). The <cluster_clause> may be missing if the <object_clause> is present.

The <object_clause> is a boolean combination of <object_group>s. The <object_group> is a group of <object_predicate>s and a <position_relation>. The <object_predicate> has the form: <object_name> <degree>. The <position_relation>s are E, W, N, S, NW, SW, NE, SE, CONTAINING, OVERLAPPING, where E, W, N, S denote the four cardinal points. The object group is interpreted from left to right. The position relation considers only the two objects which it connects. It is fulfilled if it stands for at least one couple of objects in the image.

The <object_clause> must be evaluated, according to he boolean expression, taking into account only the images in the image database, containing those objects, named as <object_name>, with < degree > value, falling into the belief interval. The WITH operator in an <object_predicate> serves the purpose of adding conditions to the attributes associated to the object named as <object_name>. The attribute predicate is a boolean combination of attribute names. In the attribute name comparison with a numerical value, by the operation <, <=, >, >=, = and with a string value, by MATCH operation is allowed.

All the stored images fulfilling the query constitute the query answer set. With this approach the query answers can be ordered by decreasing similarity to the query specification, so an user can receive a ranked output of the retrieved images. (These advantages are typical of the information retrieval techniques).

The formal description of the proposed language and the image query processing algorithm are given in [17].

2.6. Image Output and Communication through Images

After finding the images that satisfy the query specification, corresponding tools for their output (in different colours, shapes, positions, etc.) are required.

The purpose of the communication is to organise the image exchange through local area network.

3. CONCLUSION AND FUTURE WORK

IM_DBMS is under development on SUN/3 workstations, under Unix 4.2. It is being written in C and Quintus Prolog and uses the SUNCORE graphical package [19].

The design of IM_DBMS is based on the experiences of previuos prototypes. The first prototype, described in [15], was based on fuzzy set techniques. It was implemented on IBM PC/AT computer and was intended for the management of business graphical images, generated by a commercial business graphical editor (IBM Graph Assistant). Another prototype, working with graphical images, is described in [16] and is applied for house furnishing design drafts.

We plan to apply the results obtained with IM_DBMS in the project MULTOS, which is part of the ESPRIT (European Strategic PRogramme for Information Technology) [3] and in the project Multimedia information processing supported by the Bulgarian Ministery of culture, science and education. In MULTOS, a first prototype for the storage (based on optical media) and retrieval of multimedia office documents has already been implemented. However, in this prototype images are treated as passive components in the multimedia documents, that is, components which are retrieved as part of the document but cannot actively contribute in the retrieval process (no condition on images can be part of the query, only conditions on attributes, text and the document structure) [4]. In the second MULTOS prototype, we plan to build a specialized subsystem, functionally similar to IM_DBMS, which will allow a higher integration of images in the document retrieval process.

REFERENCES

[1] Ballard, D., and C. Brown: Computer Vision. Prentice Hall 1982.

[2] Barnett, J.: Computational Methods for a Mathematical Theory of Evidence. *Proc. 7-th Inter. Joint Conf. on Artificial Intelligence.* Vancouver, B.C., (1981) 868-875.

[3] Bertino, E., S. Gibbs, F. Rabitti, C. Thanos, and D. Tsichritzis: Architecture of a Multimedia Document Server. *Proc. 2nd ESPRIT Technical Week.* Brussels (1985).

[4] Bertino, E., F. Rabitti, and S. Gibbs: Query Processing in a Multimedia Document System. *ACM Trans. on Office Information Systems.* Vol. 6 (1988) 1-41.

[5] Blaser, A., (Ed.): Database Techniques for Pictorial Applications. Springer Verlag, Berlin 1980.

[6] Chang, S., and K. Fu (Ed.): Pictorial Information Systems. Springer Verlag 1980.

[7] Chang, S., and T. Kunii: Pictorial Data-Base Systems. *IEEE Computer.* Vol.14, 1981.

[8] Chang, S.: Image Information Systems. *Proc. of the IEEE.* Vol. 73, No. 4, (Apr. 1985) 754-764.

[9] Eshera, M., and K. Fu: An Image Understanding Systems Using Attributed Symbolic Representation and Inexact Graph-Matching. *IEEE Trans. Pattern Anal. Macth. Intell..* Vol. PAMI-8, No.5, (Sept. 1986) 604-618.

[10] Goos, G., and J. Hartmanis (Ed.): Pictorial Information System. Springer Verlag 1980.

[11] Gordon, J., and E. Shortliffe: The Dempster-Shafer Theory of Evidence in Rule-Based Expert Systems. In: Buchanan, B., and E. Shortliffe, (Ed.): Mycin Experiments of the Stanford Heuristic Programming Project. Addison - Wesley Publishing Company (1984) 272-292.

[12] Iyengar, S., and K. Kashyap: Guest Editors' Introduction. Image Databases. *IEEE Transaction on Software Engineering.* Vol. 14, No. 5, (May 1988).

[13] Nagy, G.: Image Database. *Image and Vision Computing.* Vol. 3, No. 3, (Aug. 1985) 111-117.

[14] Rabitti, F., and P. Stanchev: An Approach to Image Retrieval from Large Image Databases. *Proc. ACM-SIGIR Conf..* New Orleans (1987) 284-295.

[15] Rabitti, F., and P. Stanchev: Graphical Image Retrieval from Large Image Databases. *Proc. AICA Conf..* Vol. 2, Trento (1987) 69-89.

[16] Rabitti, F., and P. Stanchev: GRIM_DBMS: a Graphical Image DataBase Management System. *Proc. IPIP Working Conf. on Visual Database Systems*, Japan (1989).

[17] Rabitti, F., and P. Stanchev: IM_DBMS: Image Query Language and Image Query Processing. to be published (1989).

[18] Salton, G., and M. McGill: Introduction to Modern Information Retrieval. McGraw-Hill 1983.

[19] SunCore graphical package. *SUN 3 Reference Manuale* 1986.

[20] Tamura, H., and N. Yokoya: Image Database Systems: A Survey. *Pattern Recognition.* Vol.17, No.1 (1984) 29-43.

[21] Tsichritzis, D., and F. Lochovsky: Data Models. Prentice-Hall, Englewood Cliffs, N.J. 1982.

ZENITH - A PROCEDURAL IMAGE PROCESSING SYSTEM

V. Darie, M. Cârstea, C. Florescu, L. Berechet, C. Mecota [1)]

Abstract

ZENITH system rends possible images processing of great dimensions, stored as files on magnetic devices. There is implemented a set of classical functions used for image processing.

It is presented an original technique used in geometrical image transformations, based on control points specification.

Image format

Images used by the ZENITH system are obtained by a digitization process performed with scanner devices. These images are stored as files on magnetic devices (tapes or disks).

The image has dimensions in the range of 1 to 8000, and are coloured by 255 gray levels, quite sufficient to cover most of the situations. Images are considered to be virtual matrices on disks or raster memories, the virtualization being accomplished by some access routines through the memory buffers.

The need to access pixels of an image stored on disk is generally the same for most of the ZENITH functions, as well as for other functions that the user might define. According to this, several general purpose routines have been worked out in order to provide the access to image pixels. These routines realize: the opening and closing of an image file the reading/writing of a pixel at random, the reading/writing of a pixel sequentially.

---

1) RESEARCH INSTITUTE FOR COMPUTER TECHNIQUE AND INFORMATICS
72321 Bucharest, Calea Floreasca 167, sector 2 - ROMANIA

General functions for image processing in the ZENITH system

ZENITH implements (in a new context) some classical techniques used for image processing: smoothing, sharpening, rectangular or polygonal clipping, pixel by pixel logical or arithmetic operations, fast FOURIER and WALSH transforms, image filtering, image displaying on graphical devices, histogram modification, feature vectors classification using supervised or non-supervised techniques (see references I1I, I2I, I3I).

The functions in the ZENITH system allow rectangular windows specifying within source images as well as within the destination ones.

Being oriented to procedural image processing, the functions of the ZENITH system do not involve any dialogue with the user, the required parameters of each function being read from a previously written file. Interactive auxiliary functions have been provided for specifying the required parameters in order to create these files. The user can develop his own routines for writing (interactivly or not) parameter files, compatible with the ones required by the ZENITH functions.

Applications developed using ZENITH functions can build up new functions that will extend the already existent ZENITH functions, will be called from the catalogued procedures and will adapt image processing to each particular application.

The MIX function of the ZENITH system performs superposing several source images to get the destination image, by using a particular way of considering pixels from the images. Various operations can be performed: covering parts of an image with fragments of other images, pixel by pixel algebraic and logical operations (addition, substraction, multiplication, division, logical "not", "and", "or"). There is also available the possibility for both pixels in the source and destination images to be translated using a "look up table", when sharpening or histogram equalization is wanted.

The SPLIT function allows extracting, from a given window in the source image, one or several subimages that are stored in files with specified names. These subimages can be disjunctive or not.

POLYGON function allows several types of operations by using polygons applied to images: creating images containing filled polygonal areas or just contours of such areas; hatching the inside or outside of a polygonal area at a given gray level; drawing out of an image the inside or outside of a polygon; drawing a polygonal contour of a given breadth over a given image.

The SMOOT function performs image smoothing by replacing the gray level of each pixel within a given window in the source image by a weigthed sum of the gray levels of the neighbouring pixels.

The ZENITH system has a specified function for image sharpening, named SHARP.

The CONV function is a facility to utilize a filter between the processing of a direct and inverse WALSH or FOURIER transformation.

The ZENITH system has been provided with facilities for computing and modification of histograms.

Geometrical transformations

ZENITH system is provided with the GT function, which performs some geometrical transformation of an image into another one.

Many such functions are related to some linear transformations, such as rotations, translations, homotheties or more complex linear transforms. In such cases, three or less control points are enough to compute the six parameters of a global linear transform:

$$I = a \times I_1 + b \times J_1 + c$$
$$J = d \times I_1 + e \times J_1 + f$$

where (I_1, J_1) are the coordinates of a pixel from the transformed image and (I, J) the corresponding pixel from the source image.

A more complicated heuristic approach is used within the GT function, when the set of control points contains more than three points. In such cases the geometrical transformation is approximated by several local-linear transformations, applied to some small areas. The trouble is that boundaries between these areas (transformed by distinct local-linear transformations) are not accurate: some overlappings or non-covered areas or misconnections of details may appear.

This paper describes an heuristics for performing such a triangulation, the triangles having their vertices on the defined control points. The algorithm consists of two parts:

- covering with enough good triangles the image area included in the convex hull generated by the set of control points;
- covering with adequate triangles the rest of image untill the image boundaries.

The first part of the algorithm is based on establishing of an initial oriented polygonal simple area (which has no control points inside) and successive extending this area by adding new control points to the current polygonal line or by excluding some elementary concav triangles from the polygonal current line (including them into interior of polygonal area previously considered).

Algorithm starts from a polygonal closed line, consisting of two nearest points among control points set. Let $P_1(I_1,J_1)$ and $P_2(I_2,J_2)$ be these points. The initial closed polygonal area consists of segments P_1P_2 and P_2P_1.

Suppose that $\mathcal{A}$ is the control points set, that $\mathcal{P}$ consists of points currently on the polygonal closed line, in this succession:

$$P_1P_2P_3.....P_nP_1 ,$$

that $\mathcal{E}$ is the set of points already included in the polygonal area and $\mathcal{R}$ is the subset of $\mathcal{A}$ not included yet to $\mathcal{E}$ or $\mathcal{P}$ sets.

Thus the algorithm of selecting the elementary triangles to be added to the polygonal current area consists of several steps:

Step 1: Set the initial conditions:

$$\mathcal{E} = \{\emptyset\} , \mathcal{P} = \{P_1, P_2\}, n=2 , \mathcal{R} = \mathcal{A} \setminus \mathcal{P}$$

Step 2: Successively analyse each edge of current polygonal line, e.g. $P_oP_1, P_1P_2, ...P_nP_1$

and for each such edge consider the triangles of two types:

- of the form $P_kP_{k+1}P_{k+2}$ if from oriented edge P_kP_{k+1} the P_{k+2} vertex is seen on the right and only if there is not an another point from $\mathcal{A}$ inside this triangle (case (a), concavities);
- of the form $P_kP_xP_{k+1}$, where $P_x \in \mathcal{R}$ and P_x is seen "at the right" of the P_kP_{k+1} segment, and also this triangle does not contain a point from $\mathcal{A}$ inside it (case (b)).

For each such accepted triangle a value is computed to be minimised among whole set of accepted triangles. This value is considered (with enough good results) to be max (P_kP_x, P_xP_{k+1}) in case (b), or $P_kP_{k+2}-D_o$ in case (a).
That means that this is a min-max value. D_o is a constant used to encourage the exclusion first of the concavities from current polygonal line instead of adding new ones;
Suppose k is the selected edge corresponding to minimum.

Step 3: Accordingly to each case retained in Step 2, an action is performed:
- in case (a):
- the point P_{k+1} is excluded from $\mathcal{P}$ and included to $\mathcal{E}$;
- relatively to new edge of the polygonal line P_kP_{k+2}, P_{k+1} is memorized as the point of the elementary triangle, selected to do a local linear transform;
- $n = n - 1$
- in case (b):
- P_x is included to $\mathcal{P}$, P_kP_{k+1} is excluded and P_kP_x, P_xP_{k+1} are added to new closed polygonal line; P_{k+1} is marked for P_kP_x (and P_k for P_xP_{k+1}) as parteners to build an elementary triangle to perform local linear transform;
- $n = n+1$

Step 4: If no triangle is selected at Step 2, this means that the polygonal closed line is convex, and $\mathcal{R} = \{\emptyset\}$. Thus Step 3 is skipped and first phase of algorithm is accomplished. Otherwise (if a triangle was selected) Step 5 is performed.

Step 5: Compute coefficients of local-linear transform relatively to the selected elementary triangle and perform the transform. Then continues with Step 2.

The second part of heuristics is based on the results of the first part, that means that the final closed polygonal line:

$$P_1P_2 \ldots\ldots P_nP_1$$

must be convex and to each edge P_kP_{k+1} a point Q_k is retained as partener to the previous two, to perform an elementary triangle in order to apply a local-linear transform inside of convex polygonal line $\mathcal{P}$. The coefficients of these transforms used at internal side of polygonal oriented line will be extended to the external side of this edge to whole area delimited by the consecutive two externaly prolonged bisectrices, till these reach images limits.

Remarks

1. In the above described algorithm the succession of indices k, k+1, k+2 must be considered circularely through n, that means ...n-1,n,2. ...
2. Some optimization may be considered relatively to the first part of algorithm. For example, if Z is a partial optimum and the length of P_kP_{k+1} is greater than 2Z, then is not meaningly to test points P_x from $\mathcal{R}$. To do a more rapidly selection of a best triangle in order to extend current polygonal area $\mathcal{P}$, it is possible to take a minimum relatively to a part (which rotates on polygonal line) of polygonal line, instead of whole polygonal line.

References:

I1I. Gonzalez, R.C. and Wintz, P. - "Digital image processing" Addison-Wesley, 1981

I2I. Proc. IEEE I1972I Special issue on digital picture processing, vol. 60, no. 7

I3I. Tou., J.T., Gonzalez, R.C., "Pattern Recognition Principles", Addison-Wesley, I1974I

PORTABILITY OF IMAGE PROCESSING SOFTWARE

H. Suesse , K. Voss 1)

Since in the last years the hardware of microcomputers often changes and so also changes the environment of a software product, it is even at times necessary to transport a software product from one to another environment. The life cycle of software must be made longer; that is, it must be designed to survive hardware changes. The tendency to portable software seems to have grown with the proliferation of micro-and minicomputers. It appears that the most important matter in producing portable software is the use of a higher-level programming language but it raises a lot of problems. The portability of image processing software is a more complicated problem because the software requirements of image processing are somewhat different from those of general computing (/7/).

1. Software portability

1.1. General problems

A given program is used in a certain environment that consists mainly of two parts: the material environment (the computer and its devices) and the software environment (the operating system with many tools). Furthermore, we can distinguish two software environments of a program: the nearest neighbourhood environment(such modern systems like Turbo-Pascal,Turbo-C,Forth, and so on) and the global environment (the operating system like UNIX,MS-DOS). Therefore we have not only to make portable the program, we have also to make portable the nearest neighbour-hood environment with all tools in this environment, like compiler, editor,run time system and so on.
We distinguish two strategies of software portability:

a) Global portability
In this case either the material environment changes completely or the global software environment changes. For instance, in the source environment we have the UNIX-system and in the target environment we have the MS-DOS-system or the central processor is another one.

b) Local portability
In this case the environment only partly changes. For image processing software it often happens that only parts of hardware changes, this is for example:

Number and dimension of the image memories are different, or it is a difference in the number of gray levels, in the graphic devices, special processors, and so on.

The local portability means, that the image processing system must be generated by flexible parameters and it is to construct modular. If only parts of hardware changes, then either we have to adapt the parameters or the moduls are to interchange. But there are a lot of other problems:

----

1) Friedrich Schiller University, Dept. of Technology,
Ernst-Thaelmann-Ring 32, Jena 6900, DDR

Image processing software and their nearest neighbourhood environment involves a large number of algorithms for image processing problems. These algorithms must be described in a machine-independent way, so as to be useful in different contexts and on different machines. All computers use binary representation. One of the governing characteristics of a system is the length of a word. The variety of lengths is large - from 8 to 64 bits - and affects the method of representing values in integers and other numbers, the method of addressing data, the number of characters permissible in a word, etc. Only the simple problem of representing negative integers give different results and so on. Therefore portability implies also the portability of algorithms.

A program is portable if the effort required for its transport is much less than the effort required for its initial implementation and if it retains its initial qualities after the transport. The portability of a program can be evaluated by measuring the transport effort. This can be mathematically determined to be 100% portable (this is almost never achieved), 80% portable, and so on. In the following we want to develop a concept of portability of image processing software.

1.2. Only a higher-level programming language ?

A very simple, well-known strategy is the following:
We have to program the software system in a higher-level language, to transport the source texts to the target computer, to compile the source texts and to link the object programs. Then we have a new task.
This simple method raises a lot of problems:

a) A higher-level language likely to be available on the anticipated target machines. Only three languages currently qualify in this regard: FORTRAN, C, and PASCAL.

b) The lack of standardisation and the lack of precise definition have led to the development of many different dialects. It is almost impossible to find two different implementations of a language that are rigorously compatible, even if they run on the same computer and the same operating system. Moreover, implementors often do not even try to attain this compatibility, adding instead extensions and improvements to various implementations (/3,6/).

c) The fact of "machine independence" of a higher-level language is a fuzzy term. The machine independence means, that the higher-level language must be independent from the hardware, from the computer and from the operating system. There don't exist any language satisfied this "machine inde-pendence" completely (/3,6/).

d) In the image processing the diversity of periphal devices is often baffling. With the constructs of a higher-level language it is impossible to program the large variety of these devices. Therefore we need library subprograms, written in assemly language.

All things considered show that this simple way is not suitable to product portable image processing software.For all that, higher-level programming languages with software tools of transport are the basic tools for writers of portable software.

1.3. Tools of software transport

a) Macroprocessors,preprocessors

This tools translate a dialect to another dialect of the same language.

b) Higher-level language translators

This tools translate a higher-level language into another higher-level language, which is available on the target machine. For example the translator from FORTRAN to ALGOL 68 is described in /4/.

c) Verifiers,filters

This tools have the purpose to verify whether programs written in a programming language are conform to the rules and restrictions established in a description to make the programs portable. PFORT /5/ is a well-known filter for FORTRAN and LINT /2/ is another one for C.

d) Generators

This tools produce compilers in several phases (lexical, syntactical, semantic analysis, code optimization, code generation) to make the compilers portable. Such generators are well-known as compiler-compiler systems or compiler-generator systems.

All this tools of software transport have the common problem:
Not only a program must be portable, but also the tools of transport and the nearest neighbourhood with all tools (compiler, editor etc.) must be portable. This is our central problem of portability and this we must solve by a new idea.

2. Our idea of portability

a) The image processing system (IPS) involves all tools suited to program development (compiler, debugger, editor, file management, and so on) and the tool of software transport.

b) The complete image processing software is written in a system language S, like UNIX in C, FORTH in FORTH, AMBA in LAMBA (/7,8/), and so on. Only a minimal kernel is written in assembly language (ca. 2-20%). Only using this kernel, the tool of software transport, and the source texts, the complete system is generated on the target computer by a boot-strap-process (see figure 1).

c) We distinguish a lot of levels of the IPS, the lowest one is the kernel. A higher level is built by the lower one and a higher syntactical term of the system language is also built by lower syntactical terms. The system kernel has an interface to the other levels of the system and is the start point of the bootstrap-process. The higher levels and a little part of the kernel have to guarantee the local portability of the system.

d) The tool of software transport is a portable compiler, the so called "translator" , and differs from the compiler of the IPS. This translator is a variable code generator that has a machine-independent front end that does only semantic analysis of the system language and a code generator based on a look-up-table. The table-based code generator describes the machine's architecture and the assembly language, which is available on every target computer. The compiler of the image processing system generates an object language, that is independent of the target computer. Consequently this compiler is a "compiler-interpreter" and is itself portable. This independent object language of the compiler is the input of the translator and therefore the translator doesn't make the lexical and syntactical analysis.

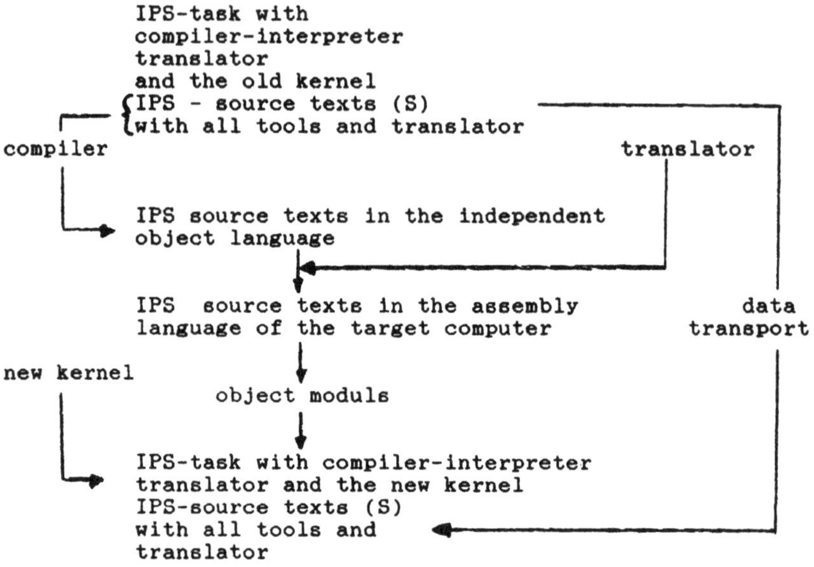

Fig. 1 - The boot-strap process

The main work for the bootstrap process is the rewriting of the minimal new kernel and of the set of tables that describe the machine´s architectur and assembly language of the target computer.

3. The portable image processing system AMBA

The idea of portability, shown in section 2, is realized by the image processing system AMBA (/7/). The system language is the so called LAMBA language. The AMBA system consists of a kernel (5%) and a lot of nonkernel tools, like:
-command interpreter (this is the highest level)
 with moduls of adaption (important for local portability),
-text processing tools
-file management
-large number of numeric software tools and tools for image processing
-compiler-interpreter
-translator.

The lowest level, the kernel, is written in assembly language. The interface with the machine´s hardware is a list of "system-calls" through which programs request services such as input-output, grafic functions, more memory, ect.
The global portability was successful realized by the boot-strap process, as shown in /1/.
The degree of portability was determined by 90% and it was typically about three months´ work for one programmer.

4. References

/1/ Hermann C., R. Neubauer, W. Ortmann, M. Schubert, H. Suesse, K. Voss: Softwaresystem AMBA auf Personalcomputer A7100. Bild und Ton 1 (1989) 11-14.

/2/ Johnson S.C., D.M. Ritchie: Unix-Time-Sharing System: portability of C-Programs and the Unix-System. Bell System Technical Journal 57 (1978) 2021-2048.

/3/ Lecarme O., M.P. Gart: Software Portability. McGraw-Hill Book Company, New York 1986.

/4/ Prudom A., M.A. Hennel: Some problems Concerning the Automatic Translation of FORTRAN to ALGOL 68. Sigplan 12 (1977) 138-143.

/5/ Ryder B.G.: The PFORT Verifier. SP-E 4 (1974) 359-377.

/6/ Schauer H., G. Barta: Konzepte der Programmiersprachen. Springer Verlag Wien New York 1986.

/7/ Voss K., P. Hufnagel, R. Klette: Interactive software systems for computer vision. In: L.N. Kanal, A. Rosenfeld (eds): Progress in Pattern Recognition. North-Holland, Amsterdam (1985) 57-78.

/8/ Voss K.: Anwendung bildanalytischer Verfahren in der Histopathologie. In: G. Burger, M. Oberholzer, W. Goessner (eds): Morphometrie in der Zyto-und Histopathologie. Springer 1988, 149-163.

AUTOMATIC IMAGE PROCESSING IN NONDESTRUCTIVE TESTING

EHLERS[1], MEYENDORF[1], MORGNER[1], POHLE[1], ECKELT[2], GRAMBOW[3], RICHTER[3]

[1] "Otto von Guericke" Technical University, PSF 124, Magdeburg 3010

[2] VEB Robotronvertrieb Berlin, PSF 1235, Berlin 1086

[3] VEB SKL Magdeburg, Alt-Salbke 6-10, Magdeburg 3012

In many spheres of production it is for safety reasons that nondestructive weld inspection is called for as weld defects are liable to affect the loading capacity of the weld. Frequently, radiographic inspection methods are used in view of their capability of providing meaningful information. Evaluation of radiographs, however, is still governed by the subjective influence of the operator.

Similarly, this problem is faced in ultrasonic weld inspection when it comes to evaluating signals and images obtained from scanning specimens or components by means of locally acting test systems. Hence, to objectify and improve inspection, automation of the testing process by use of automatic image processing has been called for [1-3].

In an approach to automatic evaluation of weld radiographs, VEB SKL Magdeburg, VEB Robotronvertrieb Berlin and Technical University of Magdeburg jointly developed the DIXI (Digital Inspection of X-Ray Images) program package for the A 6472 image processing system [4]. Adapted to perform image processing, flaw segmenting, flaw gauging and classification, DIXI comprises all the routines needed for image analysis by means of digital image processing.

Using an image processing system in the SKL company, the investigation was performed on 184 radiographs from the current production. Reflecting 440 weld defects, the radiographs were scanned by a T.V. camera and, adopting A/D conversion, an image matrix was obtained to which the following processing steps were applied:

* T.V. image cumulation to compensate for local noise of the camera:
* Pixel restoration to eliminate spot-related aberrations;
* Median filtration to further improve the signal-to-noise ratio;
* Background subtracting to eliminate the image background to the greatest extent possible;
* Binarization to localize flaws;
* Readout of the line image and its Freeman coding;
* Flaw gauging to determine quantitative defect characteristics.

In flaw gauging, the flaw contour determined and the original flaw were used to establish for each defect twenty parameters that can be utilized for classification. Once a pool of knowledge comprising for each defect class maximum, minimum and mean values as well as the standard deviation of the parameters had been established, classification was performed.

Two classification programs viz the minimum-distance classifier and the parallelepiped classifier, were adopted to distinguish all defect classes as per TGL 10646/01 standard specifications, discriminate between point defects (pores, pinholes, pipes, slag inclusions) and line-shaped defects (slag lines, lack of fusion, root defects, cracks), and classify line flaws into two-dimensional and three-dimenional defects. The table below reflects the results obtained from applying both classification methods.

DEFECT CLASSES TO BE DISTINGUISHED	EVALUATION OF RISK	MINIMUM-DISTANCE CLASSIFIER	PARALLELEPIPED CLASSIFIER
All classes	Correct	72%	52%
	Too strict	18%	2%
	Poor	10%	3%
	Unclassifiable	-	43%
Point and line defects	Correct	79%	83%
	Too strict	20%	-
	Poor	1%	2%
	Unclassifiable	-	10%
Two-dimensional and three-dimensional defects	Correct	72%	Not feasible
	Too strict	17%	
	Poor	11%	
	Unclassifiable	-	

The present findings confirm that image processing technology lends itself to automation of radiographic inspection. Prior to field use, however, further investigations and improvements of the DIXI program package will be needed.

REFERENCES

[1] Halmshaw, R.: Real time radiography - A review of recent developments. Proc. XIIth WCNDT, Vol. 1, pp. 139-154, London, 1987.

[2] Daum, W., Rose, P., Heidt, H. and Builtjes, J. W.: Automatic recognition of weld defects in X-ray inspection. Brit. J. NDT 3 (1987) 79-82.

[3] Keck, R. and Coen, G.: Automatische Auswertung von Röntgenbildern mit dem System REBUS. Qualität und Zuverlässigkeit 32 (1987) 33-36.

[4] Eckelt, B.: Anwenderdokumentation DIXI. MEC. VEB Robotronvertrieb, Berlin, 1988.

IMAGE PROCESSING OF TWO-DIMENSIONAL GEL ELECTROPHORESIS

K.-P. Pleissner

Academy of Sciences of the GDR
Central Institute of Molecular Biology

1. Introduction

Two-dimensional gel electrophoresis of proteins is a powerful technique for separation and analyzing of up to thousands of proteins particularly interesting for biological research and clinical investigation /9,1/. The 2D polyacrylamide gel electrophoresis (2D-PAGE) involves isoelectric focusing in the first dimension and sodium dodecyl sulfate (SDS) electrophoresis in the second, yielding a two-dimensional map in which each protein is separated on the basis of chemical properties (pH-values) in one direction and on the basis of molecular weight in the other. On a photographic film each radioactively labelled protein is represented by a spot, the position, the intensity and the area of which characterize the corresponding protein. The biochemical process for the preparation of a gel is rather complex and requires accurate procedures to obtain reproducible results since various nonlinear distortions may appear. Due to the great amount of information to deal with, several computerized systems for automated analysis of gel electrophoresis images have been developed (TYCHO, QUEST, GELLAB, HERMeS) /2,4,6,19/, and different image acquisition devices, computers and algorithms are used /20,7/. The crucial problems lie in the preprocessing of the images for segmentation and spot detection, matching of different gel images for a quantitative comparison of protein patterns, in the standardization and reproducibility of results and reduction of processing time. The automated analysis of electrophoretic gels contains the following tasks:

- Digital acquisition of images with corresponding spatial and densitometric resolution and storage of the digital images
- image processing, that means, reduction of the statistical noise and elimination of the streaks typical for gel images
- segmentation and detection of spots and generation of a spot list which contains the local coordinates, the intensity and area of spots
- matching of two or more electrophoretograms for a comparison of an unknown sample to a normal one or in controlling a biological process
- interpretation of the results of gel matching using, for instance, the experience of experimentators or a knowledge-based approach.

2. Image acquisition and spot detection

We use a modified image processing system BVS 6471 /5/ consisting of a 16-bit computer CM 1420 (1-MB main storage, 100-MB disk storage), a graphic control unit K 6067 with an image memory of 768 x 512 x 8 bit equipped with a color monitor and a drum scanner FEAG 200 /3/ or a CCD-camera C 222 with 8-bit video AD-converter /15/ and a lightbox for the acquisition of photographic film obtained from a gel. Using the FEAG the raster size equals 40 μm (about 25 pixel per minimum spot area) and the optical density is 2 O.D. In contrast to FEAG, the acquisition by means of a CCD-camera does not require a photographic reduction of the original size of gel and needs much less time for acquisition but the demanded spatial resolution is ensured only in a small area of gel (Fig.1). A comparison between original and scanned image can be done visually using the monitor of the graphic control unit. For precise acquisition the FEAG 200 is preferred. The digital images are stored on disk. For noise reduction a smoothing procedure of the grey levels was performed using a 5 x 5 separable low-pass digital filter with coefficients (1 2 3 2 1) per line. Because of the inhomogeneities of the background and because the intensity between two spots often did not reach the background level, a special algorithm for spot detection developed by Zimmer and Neuhoff /21/ and described in detail by Prehm /12/ is used. The procedure is based mainly on the detection of spots by the evaluation of the curvature of the grey level surface. For each pixel the curvature is evaluated in a 9 x 9 neighborhood using a 2-dimensional band-pass filter (or convex area detector) with coefficients shown in Fig.2. If the curvature was greater than zero, the central point belonged to a spot kernel, otherwise to the background. If the spots are located extremely close or may occur as a "shoulder", these spots were often not detected as two different spot kernels since the area of the grey level surface between the spots did not show a negative curvature. However, the degree of curvature in these areas was lower than in the area of the spots. For separation of these closely located spots a surrounding area with 7 x 7 pixel was used and the difference in the degree of curvature between the central point (C) and the pixel of the rim (R) was calculated. Pixel C was attributed to the background if at least for one of four directions (Fig.3) the two differences between the degrees of curvature $|K(C) - K(R)|$ were smaller than a fixed threshold P1. In order to suppress spots with very small area, an additional threshold P2 was introduced. All pixels of a spot kernel whose area was smaller than P2 were attributed to background. The output of the spot detection algorithm is a spot list with running number of the spots, their x-y coordinates, their integral, minimum/maximum optical density value and their area (Fig.4). For a 512x512 image the operating time for spot list generation on our machine takes about 15 min. A problem of such spot detection procedure is that the number of spots depends on the threshold parameters P1 and P2. Comparisons between automatically detected and visible spots done by Prehm /12/ recommend P1=1 and P2=40.

3. Gel matching

In order to control a biological process or to examine the metabolic activity of a cell under different sets of experimental and environmental conditions, a comparison of different electrophoresis images is necessary. The so-called gel matching on the basis of the spot list is the most difficult task of the gel image processing because the geometry of gel images are globally and locally distorted due to the inhomogeneities of polymerization and due to instability of electrophoretic conditions. There exists a lot of algorithms for gel matching based on trigonometric /2/, probability-theoretical /8/, graph-theoretical /17/, syntactical /18/, and other approaches, and matching can be done in a "hand-matching", semi-automatic or automatic manner in dependence on the "region of interest" of gels, the number of spots to be analyzed and the possibility of man-machine interaction. Unfortunatly, a comparison and assessment of the existing gel matching algorithms were not published. But for all algorithms it is common that one has to find the corresponding point pattern in two images and to make a geometric transformation of the first image into the second one. If the analysis system includes computer-graphics one can simultaneously overlay the two point patterns after transformation and estimate which spots correspond to each other or which spots are lost or added. In our study we investigated a "hand matching" and an automatic approach for finding the corresponding "landmark"-spots in each gel.

3.1. Hand matching approach

Although the hand matching is relativly time-consuming, the recognition power of man and the occurrence of interactive computer graphics cause us to use this approach under our condition, especially if the number of proteins investigated is comparativly small. Using the image-processing software system AMBA /14/ it is possible to pick out the corresponding landmark-spots of two images by the trackball and to mark the spots of the first image, for instance, in red color and the spots of the second image in blue color. The coordinates of three spots serve as vertices of a triangle in the first image and second image, respectively. The corresponding triangles are also marked by color and the experimentator can judge whether the alignment is errorless or not, and he can repeat the hand matching once more. In this way one can split the image into several triangles and to perform an affine tranformation (scaling, rotation, translation) which maps each triangle T_j to a triangle T_i with points (Q_1,Q_2,Q_3) and (P_1,P_2,P_3), respectively /9/. In vector-space form

$$P_k = A\,Q_k \qquad k = 1,2,3$$

where $P_k = (X_{pk},Y_{pk})^T$, $\quad Q_k = (X_{qk},Y_{qk},1)^T$

and

$$A = \begin{bmatrix} a & b & c \\ d & e & f \end{bmatrix}$$

The elements of affine tranformation matrix A can be obtained as

$$\begin{bmatrix} a \\ b \\ c \end{bmatrix} = \begin{bmatrix} X_{q1} & Y_{q1} & 1 \\ X_{q2} & Y_{q2} & 1 \\ X_{q3} & Y_{q3} & 1 \end{bmatrix}^{-1} \begin{bmatrix} X_{p1} \\ X_{p2} \\ X_{p3} \end{bmatrix}$$

$$\begin{bmatrix} d \\ e \\ f \end{bmatrix} = \begin{bmatrix} X_{q1} & Y_{q1} & 1 \\ X_{q2} & Y_{q2} & 1 \\ X_{q3} & Y_{q3} & 1 \end{bmatrix}^{-1} \begin{bmatrix} Y_{p1} \\ Y_{p2} \\ Y_{p3} \end{bmatrix}$$

Because the affine transformation is only valid for spots which are within the chosen triangle, one has to prove this condition. Therefore we check the position of a point with respect to a polygon (triangle) by proving the signs of a set of three inequalities performed by triangle T_j where the vertices of this special kind of polygon are ordered in a clockwise fashion/11/. The inequalities are:

$$X(Y_{q1} - Y_{q2}) + Y(X_{q2} - X_{q1}) + (X_{q1}Y_{q2} - Y_{q1}X_{q2}) < 0$$
$$X(Y_{q2} - Y_{q3}) + Y(X_{q3} - X_{q2}) + (X_{q2}Y_{q3} - Y_{q2}X_{q3}) < 0$$
$$X(Y_{q3} - Y_{q1}) + Y(X_{q1} - X_{q3}) + (X_{q3}Y_{q1} - Y_{q3}X_{q1}) < 0$$

where X,Y - the coordinates of a spot investigated.
Connecting the original spots within the triangle T_j and the transformed spots a "vector-field" is calculated in which one can evaluate whether the spots correspond to each other or not (Fig.5).

3.2. Automatic matching

For automatic finding of "lankmark-spots" we use an algorithm proposed by Richter /13/ and which is based on the invariant description of the neighborhood of point patterns. This approach is invariant under geometrical transformations, such as translation, rotation, scaling and reflection and tolerant of contamination, e.g. nonsensitive to additions and deletions of spots. Each spot P_i investigated performs triangles with all pairs of points in its neighborhood and these triangles are described by its angles α and β (Fig. 6). All pairs of angles are the description of the neighborhood of point P_i. If points in the second gel image are added or deleted, then α-β pairs which cannot be identifield build up. But a set of α-β pairs remains which still enables an identification. For the definition of neighborhoods we use the Voronoi neighbors of each P_i, and the pairs are estimated using the Delaunay - triangulation (Fig.7). In order to detect the corresponding point pattern of the first gel image in the second one, the pattern of the point P_i in the first image has to be compared with all patterns of the second image. In the case of greatest accordance of pairs, the spot corresponding to P_i is matched and can be taken as landmark-spot for the following triangulation and transformation described in Sec. 3.1. The automatic matching based on the invariant description of neighborhood of point patterns needs further investigation because a sensible measure of the correspondence or correlation of patterns is not yet found.

Literature:

/1/ Anderson,N.G.,Anderson,L.: The Human Protein Index, Clinical Chemistry 28(1982)4,739-748

/2/ Anderson,N.L.,Taylor,J.,Scandora,A.E.,Coulter,B.P.,Anderson,N.G.: The TYCHO System for Computer Analysis of Two-Dimensional Gel Electrophoresis Patterns, Clinical Chemistry 27(1981)11,1807-1820

/3/ FEAG200: Prospektmaterial VEB Carl Zeiss Jena

/4/ Garrels,J.I.,Farrar,J.T.,Burwell,C.B.: The QUEST System for Computer-Analyzed Two-Dimensional Electrophoresis of Proteins, in: Two-Dimension Gel Elektrophoresis of Proteins, J.E.Celis, R. Bravo, ed. pp. 37-91, Academic Press, London 1984

/5/ Kempe,V. et al: The interactive image processing console A6471.-Proc.6-th ICPR,Munich 1982,607-609

/6/ Lemkin,P.F.,Lipkin,L.E.: GELLAB: A Computer System for 2-D-Gel Electrophoresis Analysis I. Segmentation of Spots and System Prelimaries Computers and Biomedical Research 14(1981)3, 272-297

/7/ Mann R.-C., Mansfield B.U., Selkirk J.V.: Automated analysis of digital images generated by two-dim. gel electrophoresis Pattern Recognition in Practice II, p. 301-312

/8/ Miller,M.J., Olson,A.D., Thorgeisson,S.S.: Computer analysis of two-dimensional gels. Automatic matching, Electrophoresis (1984)5,297-303

/9/ O'Farrel,P.H.: High resolution two-dimensional electrophoresis of proteins, J.Biol.Chem. 250(1975)4007-4021

/10/ Ogawa,H.: Labeled point pattern matching by Delaunay triangulation and maximal cliques, Pattern Recognition 19(1986)35-40

/11/ Pavlidis,T.: Algorithms for graphics and image processing,Computer Science Press Inc.,1982

/12/ Prehm J., Jungblut P., Klose J.: Analysis of two-dimensional electrophoretic protein patterns using a video camera and a computer - II. Adaption of automatic spot detection to visual evaluation, Electrophoresis (1987)8, 562-572

/13/ Richter,G.M.: Identifizierung von Objekten in Bildfolgen durch invariante Beschreibung der Umgebung, Studientexte Digitale Bildverarbeitung Heft 93/1986,S.102-107,TU Dresden

/14/ Roth,K. et al: Hardware und Software des Bildverarbeitungssystems A6471-AMBA/R, BILD UND TON 37(1984)358-361

/15/ Seifert,G., Pleissner,K.-P., Lucius,H., Anders,W., Beuthan,J.: Einsatz der digitalen Bildspeicherung beim CCD-Roentgenfernsehen, Feingeraetetechnik 37(1988)3, 118-122

/16/ Serpico S.B., Vernazza G.: Problems and prospects in image processing of two-dimensional gel electrophoresis, Optical Engineering 26(1987)7, 661-668

/17/ Skolnick M., v. Neel J.: An Algorithm for Comparing Two-Dimensional Electrophoretic Gels with Particular Reference to the Study of Mutation, in: Advances in Human Genetics, vol. 15 ed. by H. Harris (Plenum Publishing Corporation, 1986)

/18/ Vincens,P. Paris,N.: HERMeS:A second approach to automatic analysis of two-dimensional electrophoresis (Part 3,Spot List Matching), Electrophoresis (1987)8,100-107
/19/ Vincens,P.,Paris,N.Pujal,J.L. et al: HERMeS: A second generation approach to automatic analysis of two-dimensional electrophoresis (Part I:Data acquisition), Electrophoresis (1986)7,347-356
/20/ Vo,K.Ph.: Computer Analysis of Two-Dimensional Gels, Analytical Biochemistry,112(1981) 258-271
/21/ Zimmer,H.G.,Neuhoff.V.: Quantitative Auswertung von zweidimensionalen Mikro-Duennschicht-Chromatogrammen, in:Informatik-Fachberichte 8, H.Nagel Ed.,Springer-Verlag,Berlin-Heidelberg-NewYork 1977,S.12-20

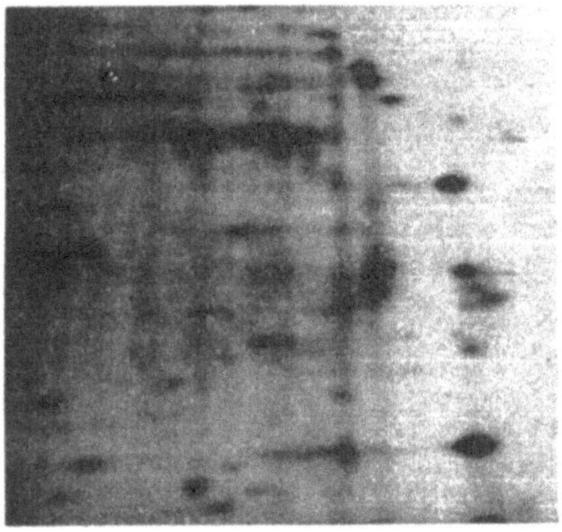

Fig. 1 Electrophoretic gel image scanned by CCD-camera

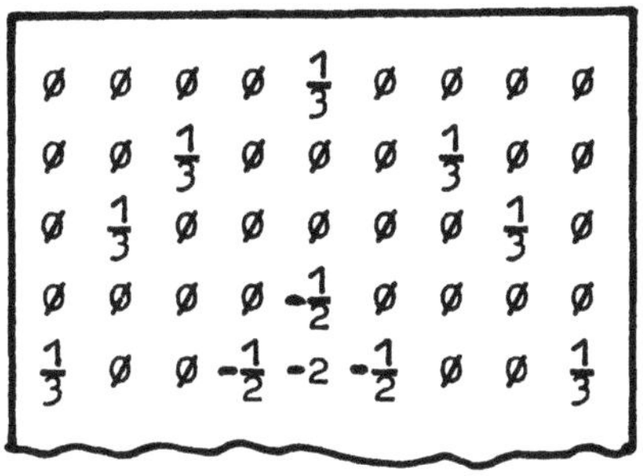

Fig. 2 Digital band-pass filter for detection of convex area (upper part)

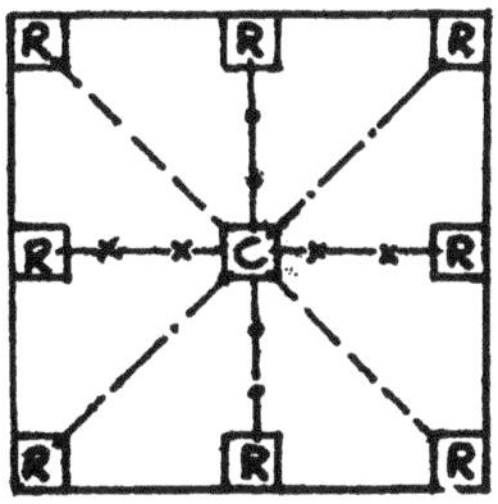

Fig. 3 Template for detection of minimum of convexity in a spot kernel (4 directions) C-center, R-rim

S P O T L I S T

Nr.	X	Y	INT. ABSOR.	MIN. ABSOR.	MAX. ABSOR.	AREA
1	5	89	8.394	0.593	0.751	2.
2	5	58	20.439	0.550	0.758	4.
3	84	239	23.780	0.470	0.636	5.
4	39	162	33.961	0.492	0.647	8.
5	5	131	8.672	0.589	0.708	2.
6	8	246	84.913	0.580	0.980	7.
7	34	32	7.025	0.537	0.617	3.
8	34	13	8.600	0.529	0.607	3.
9	82	34	38.057	0.481	0.778	6.

Fig. 4 Part of a spot list generated by described algorithm

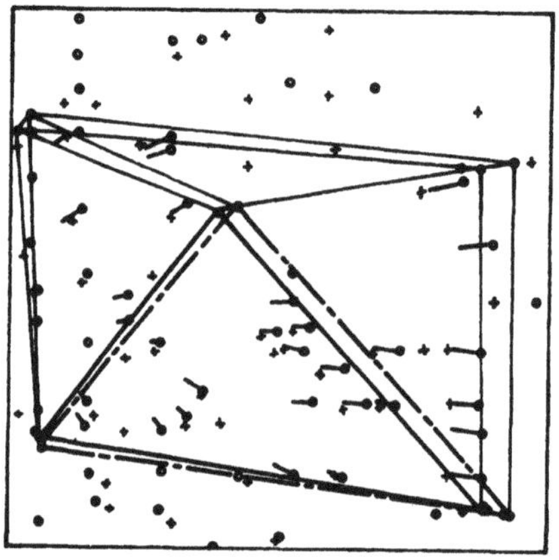

Fig.5 Results of the hand matching procedure applied to two gel images - vector field (O -spots of the first image, + -spots of the second image)

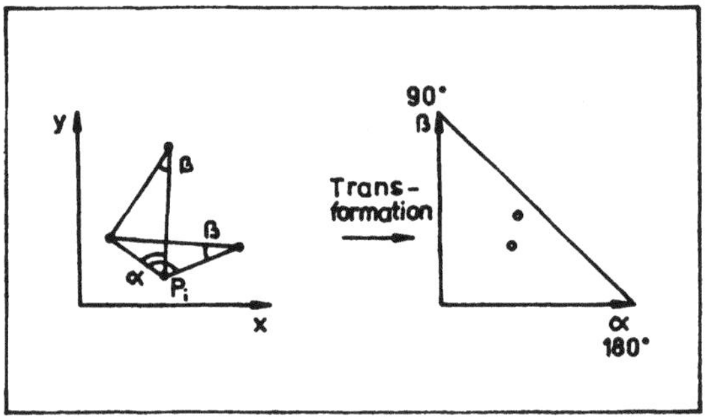

Fig.6 Invariant description of the neighbourhood of point P_i by α-β pairs

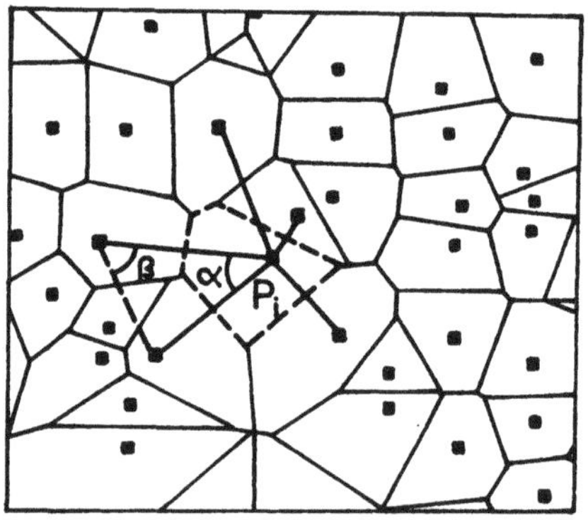

Fig.7 Voronoi tessellation of a gel image and Delaunay triangulation for point P_i (Part of gel image)

Cyclic Recognition of Simple Industrial Scenes Composed of CAD-Objects

Wolfgang Coy and Ulf Hönisch *

1. Introduction

An object recognition and scene analysis system is presented which recognizes simple industrial scenes in a cyclic, but finite process from a CAD data base. A scene will be called *simple* if it is composed of well-known objects (possibly occluded) and the composition rules are well-known. It is called an *industrial scene* if all objects are described in some CAD-data base.

A basic approach to scene analysis is to segment the (2D-)image S of the scene into objects $\sigma_1,\ldots,\sigma_n$, recognize the objects σ_1 from geometric or textural features $\varphi_1,\ldots,\varphi_t$ and (re-)-construct the —idealized— scene S' (see e.g. [7]). The result of this *data-driven* reconstruction of an image object σ may be considered to be a model μ of this object. Classification of similar image objects $\sigma_1,\ldots,\sigma_r$ consists in the match of all these objects to the same model class μ. An advantage of the data-driven analysis approach $P{:}S\rightarrow S'$ may be seen in the potentially unrestricted variety of recognizable objects (and constructable model classes), but basic recognition and analysis algorithms may become very complex especially under the practical constraints of low image resolution and the presence of noise.

Many industrial scene analysis applications differ essentially from this general problem, because the expected objects are well-known and described. This allows the use of *model-based* recognition strategies $Q{:}S\rightarrow S'$ [5,9,15], where a scene object σ_1 in the image scene S is verified to be as close as expected to some modeled object μ_i from some model base $M=\{\mu_1,\ldots,\mu_p\}$, i.e. the modeled object features $\varphi_k(\mu_i)$ are proved to match the features $\varphi_k(\sigma)$ of the scene object. The model-based analysis process substitutes the computational complexity of data-driven approaches with search processes. The verification process Q consists of a search through the model base M and comparison of the models μ with the image objects σ until the image objects are matched to some $\mu_i \in M$ (or no recognition seems to be possible).

• Model descriptions are explicit and logically (and usually physically) separated from the analysis algorithm. This property makes them exchangable.

• Model descriptions must allow the unique identification of the described objects; they should allow the extension of the model base while retaining this uniqueness property.

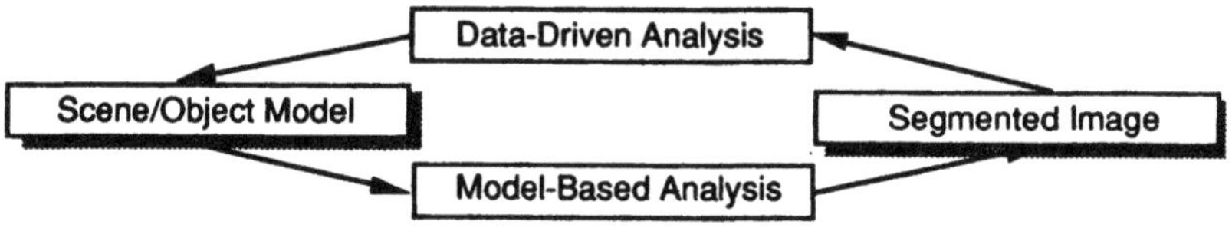

Fig. 1. Data driven and model-based analysis.

A *pure* model-based recognition system may lead from the model description to the the pixels

----

* Universität Bremen Informatik, P.O.Box 330440, D-2800 Bremen 33, Federal Republic of Germany

of the digitized image. This may be a very time-consuming process. It seems to be more interesting to reconstruct only some essential parts of the image from the model, while hypotheses about basic objects in the scene are derived in a data-driven process. The features of these partially reconstructed objects should then be verified in the captured image. In this *mixed mode* recognition process model descriptions influence the analysis process, selection of procedures to be applied, selection of parameters etc. in practically all stages of the recognition process. While date-driven recognition systems construct models, model-based systems generate a compatible set of data. In mixed mode both directions are investigated to reconstruct feature descriptions from data and from the model.

2. CAD Data Bases as Model Bases for Industrial Vision

CAD-models are considered to be good candidates for models [6], as they are independent from analysis algorithms and as they possess the uniqueness property to a very high degree (but cp. [12,13]). CAD-models are already present in the industrial design and construction processes and describe basic geometric features, which may be easily extracted from the image data [2,3,15]. DXF-CAD-models,like all early CAD-models simulating the manual industrial drawing process, are essentially 2D-descriptions of 3D-objects, where the drawing is composed of a list of primitive drawing objects like lines, circles, rectangles and arcs and their relative positions. 2D-views may be used by human users to reconstruct some mental 3D-image from these 2D-descriptions. This process is difficult to copy as a program, as the reconstruction of solid bodies from 2D-views assumes some hidden assumptions (cp. [12,13]). This situation is avoided with true geometric 3D-models, where 2D-views from 3D solid models may be automatically extracted [1,2,11].

In the actual system we use gray scale scene images of 512x479 pixels with objects described as CAD-data in DXF-format (generated by a PC-based AutoCAD-system). The scene is composed of few relatively flat but possibly overlapping objects. For demonstration purposes we used flat front panels of 19"cabinets.We assume that relatively few objects are laid down on a table of 60cm x 90cm and that camera is placed relatively high above this scene frame. This excludes many visual effects which must be treated otherwise.

3. Cyclic Approximation of Scenes as a Process of Hypothesis Verification

Three steps in the analysis cycle for simple scenes are used:

- Separation of occluded objects, Segmentation of the basic objects and initial guess.
- A repeated sequence of verification of the guesses by the use of the (stripped down) geometrical CAD-data and the exploration of alternative hypotheses.
- A decision when to stop the approximation.

Mixed-mode object recognition depends on a model-base M (here derived from a CAD-data base) and a scene image S (which is a digitized 2D-gray scale image in this case). From the image S simple features φ of objects σ_i like edges, areas and others (e.g. junctions of lines) are derived. From these feature descriptions F_i candidate models are chosen as model hypotheses $h_i \in M$ to be verified as a match in the captured image object σ_i.

There is some ingenuity necessary to find an appropriate initial guess, which distinguishes good object candidates in the CAD-based data. In the 3D-analysis, junction labels [16] and the approximate size of the objects (computed from a singular view!) are considered to result in good initial feature. As the actual implementation considers only 2D-objects a simpler hypothesis construction is sufficient with the features size, length of straight lines and the structure of parallel lines. Positional hypothesis are derived from a match process where lines of the hypothesis model are related to lines extracted form the image. Of course, overlapping objects introduce further complications, but the assumption of a finite model base reduces this problem to become manageable. Thorough examination of the features may reduce the hypothesis space drastically. A heuristic quality measure function allows the reduction of the hypothesis space, but this may introduce failures.

With the initial hypotheses space a cyclic verification process is started. Hypothesis verification is done as a model-based step, where m_i must be transformed to a constructed view v_i of the model, which should match position and orientations of the image object. This view v_i is corrected with respect to the camera parameters and is projected into the captured image S for hypotheses verification. The construction of views may become a time-consuming process; it becomes considerably faster if a number of views has been computed in advance. This should be done at least for qualitatively different views (which differ in qualitative aspects like the number of hidden lines).

Verification is successful if a number of cues in the hypothesis matches the image S (within some tolerance measure). In many applications it suffices to match only few properties from the derived hypothesis with the image. This may result in a fast analysis process. If the verification process fails it may result either in the deletion of the model hypothesis h_i or in a new view v_i' with better position and orientation parameters. In both cases a new cycle is entered (unless the number of cycles exceeds some chosen maximum).

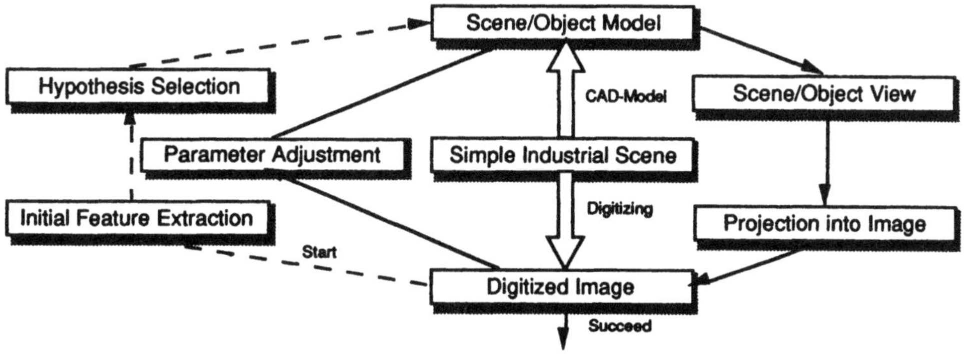

Fig. 2 A cyclic analysis model.

4. Simple 2D-Scenes

This paper reports an ongoing development. While the final goal is the analysis of three-scenes, we reduce the model-based analysis cycle for demonstration purposes to scenes composed of two-dimensional objects.

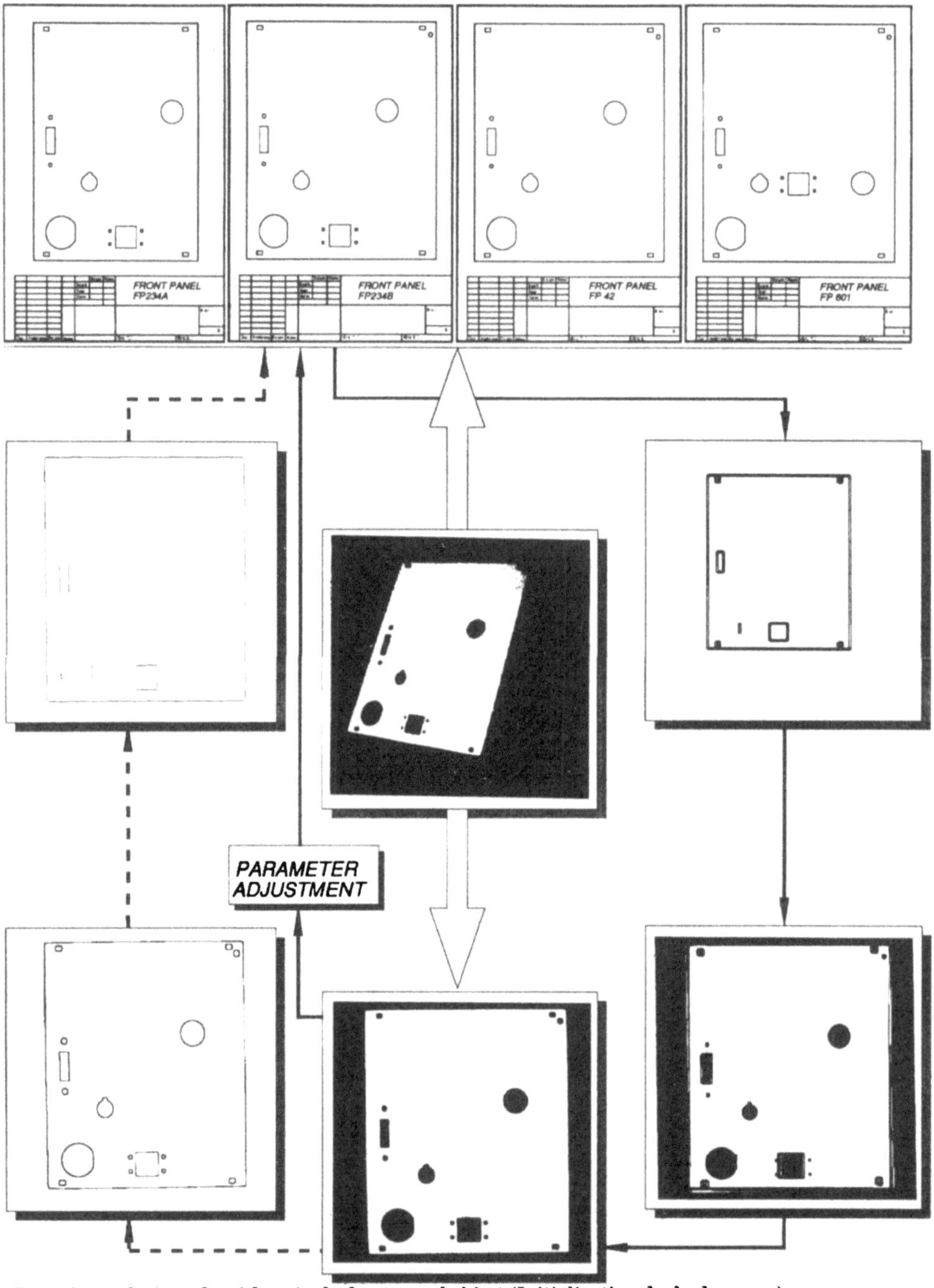

Fig.3. An analysis cycle with a single front panel object (Initialization dashed arrows).

The goal is the identification of objects and their absolute and relative positions. As the final goal is the analysis of 3D-scenes, the two-dimensional case is regarded as a specialization of the three-dimensional case. The first step consists of the separation of possibly occluded objects and the segmentation of single objects by edge detection. Then the further recognition process consists of a sequence of hypothesis generation and verification.

We describe a concrete cycle for the recognition of 19" front panels. The panels differ either in size or in the number or position of punched holes. The camera captures a 2D-gray scale picture. Overlapped panels are separated and minimally occluded parts reconstructed (as possible). After an initial edge extraction process, only straight line segments are identified [4]; this is sufficient for the ordered selection of CAD-model candidates. The most probable model is chosen and transformed to a view which fits the coordinates of the captured image. This view is projected to the digitized gray-scale image and a verification of the straight line segments is tried. Verification is done with gray-scale profiles which are investigated in orthogonal direction to the element orientation. The result of the verification is measured as the deviation of line end points; this allows a further adjustment of the transformation matrix. If the verification fails, a further cycle with adjusted transformation parameters may be performed. If the actual computation is worse than an earlier or if a maximum number of iterations is exceeded the next candidate is tried. This is repeated until hypothesis verification succeeds or fails.

The process works well in the demonstrated 2D-case with slightly overlapped objects (where occluded parts may be easily reconstructed). The actual implementation will be extended to objects with real three-dimensional properties. This complicates the hypothesis generation because the number of positional views is grossly enlarged. It demands complicated transformations of separate 2D-CAD-data as with DXF-data. Alternatively solid 3D-CAD-modeling should be considered.

Occlusion [8] is actually treated as a separate process, because it is assumed to be relatively simple in the actual context. As it may result in a series of hypotheses and falsifications, it should be included in the recognition cycle.

The data-driven part of the cyclic process could be supported by the advance generation of 'typical' views, which differ qualitatively from each other (e.g. number of hidden lines or areas).

Conclusions

The basic idea of a cyclic hypothesis/verification process proved to be sound. A mixed approach of data-driven initialization and model-based verification was shown to work in reasonable time for simple industrial scenes.

Verification of modelled properties in a captured image is an easy and reliable analysis process as compared to the complete reconstruction of a scene from image data. It shows a basic advantage, when image cues of very different size (e.g. very large and very small rectangles) are crucial for the object identification, because verification of such cues is much easier than unspecific feature extraction.

In applications, where few model properties (like some straight lines) suffice for a unique identification, the hypothesis/verification cycle may become a very fast scene analysis tool. On the other hand it is probably not very useful for general 'outdoor' scene analysis and other military applications.

References

[1] Aldefeld, B.:On Automatic Recognition of 3D Structures from 2D Representations, *Computer-Aided Design*, Vol. 15,No. 2, March 1983, pp 59-64.

[2] Bhanu, B., CAD-based Robot Vision, Guest Editor's Introduction, *IEEE Computer*, Vol.20, No.8, 1987, pp. 12-18.

[3] Bhanu, B., CAD-Based 3D Object Recognition for Robot Vision, *IEEE Computer*, Vol.20, No.8, 1987, pp. 19-36.

[4] Bruns, J. B., Hanson, A. R. & Riseman, E., Extracting Straight Lines,*IEEE Transactions on Pattern Analysis and Machine Intelligence*,Vol. 8, No. 4, July 1988, pp. 425-455.

[5] Cin,R.T. & Ch.R.Dyer, Model-Based Recognition in Robot Vision, Computing Surveys, Vol. 8:1, March 1986, pp. 68-108.

[6] Gunnarson, K. T. & Prinz, F. B., CAD Model-Based Localization of Parts in Manufacturing, *IEEE Computer*, Vol 20, No. 8, 1987, pp. 66-73.

[7] Hanson,A.R. & E. M. Riseman, Visions: A Computer System for Interpreting Scenes, in: A.Hanson and E. Riseman (Eds.), *Computer Vision Systems* , Academic Press, New York, 1978.

[8] Horaud, Radu & Skordas, Thomas, Model-Based Strategy Planning for Recognition Partially Occluded, Parts, *IEEE Computer*, Vol. 20, No. 8, 1987, pp. 58-65.

[9] Kanade,T., Region Segmentation: Signal vs. Semantics, Proc. 4th Int'l Joint Conf. on Pattern Recognition, 1978, pp. 95-105.

[10] Klette, R., Expert systems — A Modern Software Technology, Proc. CAIP '87, Akademie, Berlin(DDR), 1987, pp. 121-132.

[11] Preiss, K., Constructing the Solid Representation from Engineering Projections, *Computers & Graphics*, Vol. 8, No. 4, 1984, pp. 381-389.

[12] Requicha, A.A.G. & H.B.Voelcker, Solid Modeling: A Historical Summary and Contemporary Assessment., *IEEE Computer, Graphics & Applications* , 2:2, March 1982, pp. 9-24.

[13] Requicha, A.A.G. & H.B.Voelcker Solid Modeling: Current Status and Research Directions., *IEEE Computer, Graphics & Applications* 3: 7, Oct. 1983, pp. 25-37.

[14] Shapiro, L. G., A CAD-Model-Based System for Object Localization, Proc. SPIE 88, 1988.

[15] Shirai, Y., K.Koshikawa, M.Oshima & K.Ikeuchi, Application of 3-D Models to Computer Vision, *Computer & Graphics*, Vol. 7, No 3-4,1983, pp. 269-275.

[16] Waltz, D.I., Generating Semantic Descriptions from Drawings of Scenes with Shadows, Dissertation, AI Lab, MIT, 1972.

EXTRACTION OF TERRITORIAL DATA FROM CARTOGRAPHICAL DRAWINGS

David Hrycej, Stanislav Řízek [1)]

1. Introduction

Processing of remote sensed images of the Earth usually requires some supporting data which can be gained from other sources in advance. Territorial supporting data describe temporarily non-variable topological properties of a terrain. They can be extracted from maps and stored in a database.

Every map sheet is represented by a set of pen drawings which separate the total territorial information into the following thematic layers:

- the forests layer
- the waters layer
- the agriculture land layer
- the planimetry layer which consists of a mixture of roads, built-up areas, railways, high voltage lines etc.

These drawings contain special marks, lines and alphanumerical description which do not provide any topological information.

Processing of cartographical data consists of digitization of pen drawings, geometrical transformation, extraction of topological objects, coding resulting data and storing them in the database. Most of these tasks can be solved only by man-machine interaction.

2. Geometrical transformation

Each quadrilateral pen drawing has been digitized in an individual local coordinate system (x,y) according to Fig.1. It is to be mapped into the corresponding quadrilateral in the unique coordinate system (r,s), which corresponds to the database (see Fig.2). Bilinear functions can map exactly the corners of quadrilaterals [1], but they can cause great deformations when the corners are not read accurately. Therefore, linear transformations have been preferred.

---

1) General Computing Centre, Czechoslovak Academy of Sciences
Pod vodárenskou věží 2, 182 07 Praha 8, P.O.B. 5, ČSSR.

The following linear functions result in the minimal mean square error of mapping of four corners:

$$X = a R - b S \qquad\qquad Y = b R + a S \tag{2.1}$$

$$a = \frac{\sum_{i=1}^{4} (R_i X_i + S_i Y_i)}{\sum_{i=1}^{4} (R_i^2 + S_i^2)} \qquad\qquad b = \frac{\sum_{i=1}^{4} (R_i Y_i - S_i X_i)}{\sum_{i=1}^{4} (R_i^2 + S_i^2)} \tag{2.2}$$

where $X = x-x_0$, $Y = y-y_0$, $R = r-r_0$, $S = s-s_0$;
x_i, y_i, r_i, s_i are coordinates of the corners;
x_0, y_0, r_0, s_0 are coordinates of their centers of gravity.

Linear transformations give acceptable results. However, distortion of border regions of the quadrilaterals can cause a serious deformation of geometrical objects, e.g. a disconnection of thin curves passing through boundaries. Therefore the border regions should be pre-processed by filtering and/or by morphological transformations. Otherwise, the extraction of geometrical objects in these regions must be performed according to more complicated and sophisticated algorithms.

3. Extraction of territorial elements

The next processing step is transformation of each thematic layer into the form, which is convenient for storing it in the database. Because of binary character of pen drawings, binary morphological transformations [2] have been used. Several algorithms have been proposed and tested. They consist of elementary set operations (intersection, union, complement) and morphological operations: erosion ($X \ominus S$), dilation ($X \oplus S$), conditional dilation ($X \oplus S \,;\, Y$), homotopic thinning ($X \circ N$), branch removing ($X \circ E$) and binary median b(X) in the window 3×3 pixels; where X is the image to be transformed, S is a basic structuring element for square grid (square 3×3 pixels), N resp. E is a set of structuring elements for homotopic thinning resp. branch removing.

The forests layer is the simplest thematic layer. Forests are represented by black and dark grey regions. Binarization of images (by thresholding) results in a weak noise especially near the region borders. Reliable noise removing was performed by the binary median transformation.

Waters layer contains river lines and lake areas. The aim of the processing of this layer is:

- to remove arrow-heads along river lines
- to connect interrupted river lines (caused by bridges)
- to transform river lines onto unitary thickness.

Processing of waters layer includes separation of lake areas from river lines (opening) and the following processing of river lines: dilation (connection of interrupted river lines and merging these lines with arrows-heads), homotopic thinning (transformation river lines onto unitary thickness), short line removing and conditional dilation (removing small noise particles):

$$Y_1 = (X \oplus iS) \circ (N) \qquad (3.1)$$

$$Y_2 = (Y_1 \circ jE) \oplus (S) \; ; \; Y_1 \qquad (3.2)$$

where i, j are integer parameters (a number of iterations of an individual transformation) and ( ) denotes infinite iteration of specified transformation. Sophisticated algorithm known as pruning [2] (which removes short branches along river lines) can then be applied - if necessary - on resulting image.

The aim of processing the agriculture land layer (Fig.3) is to delete border lines of agriculture cadastres (thinner lines) and alphanumerical description of fields. Differences in the line thickness enables to delete undesirable objects by a simple transformation (binary median, homotopic thinning and branch removing):

$$Y = [b(X) \circ (N)] \circ (E) \qquad (3.3)$$

The result is thinned borders of agriculture fields (Fig.4). If it is necessary (for storing in the database), the interior of fields can be filled by conditional dilation.

The planimetry layer is the most complex one because it contains most of the information. The density of lines, marks and alphanumerical description is very high especially in the built-up areas. The lines in the pen drawings are of different thickness, but these differences are too small to appear in the digital picture. Several sequences of morphological transformations were tested with the conclusion, that it is impossible to separate distinct territorial elements only by morphological transformations. The algorithms that can at least partly separate distinct elements must utilize a prior information about their character (e.g. a road is represented by two parallel lines).

Algorithms described above depend on some integer parameters (such as a number of steps of an individual transformation). Processing of different map sheet can require slight changes in the value of these parameters.

4. Conclusion

Presented algorithms have been tested on the real data according to Fig. 2 [3]. The experimental area of about 75 km^2 has been mapped in the scale 1:25000 on 10 map sheets. Each digitized pen drawing contains 1650×2050 pixels (step 100μm). Processing of the window 512×512 pixels presented in Fig.3 and 4 took several dozens of minutes on CPU like PDP-11.

Algorithms presented in section 2 can be used for geometrical transformation of digitized pen drawings into coordinate system of database. The algorithms presented in the section 3 can be used for extraction of territorial elements without serious difficulties only in the case of simple thematic layers (forests, waters and agriculture land layer). The planimetry layer is too complicated to be processed completely by a computer. Areas, where the density of objects, descriptions and line crossings is particularly high, require man-machine interaction. Another possibility of enhancement of territorial elements extraction (which is under consideration) is to utilize information about relations between territorial elements from different thematic layers. These relations have to reflect reality; e.g. both lake and forest cannot be present in the same pixel.

5. Literature

[1] Rosenfeld ,A., A. C. Kak: Digital Picture Processing, Vol. 1 and 2, Academic Press, 1982.

[2] Serra J.: Image Analysis and Mathematical Morphology, Academic Press, 1982.

[3] Hrycej, D., S. Řízek: Programs for Processing A prior Supporting Data for Remote Earth Sensing (in Czech), Res. Rep. V-382, General Computing Centre, Prague, 1988.

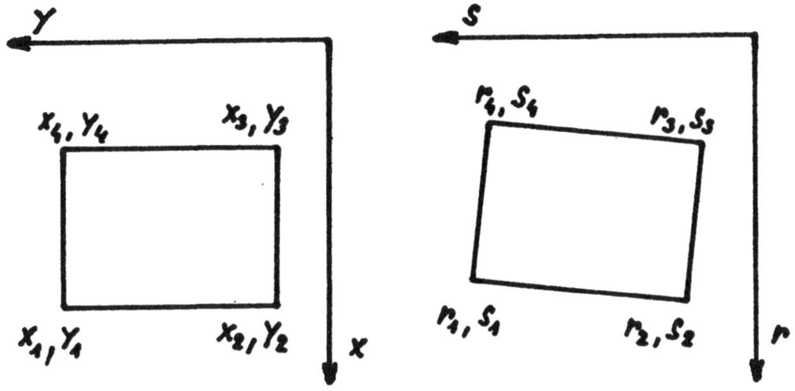

Fig.1: The coordinate systems

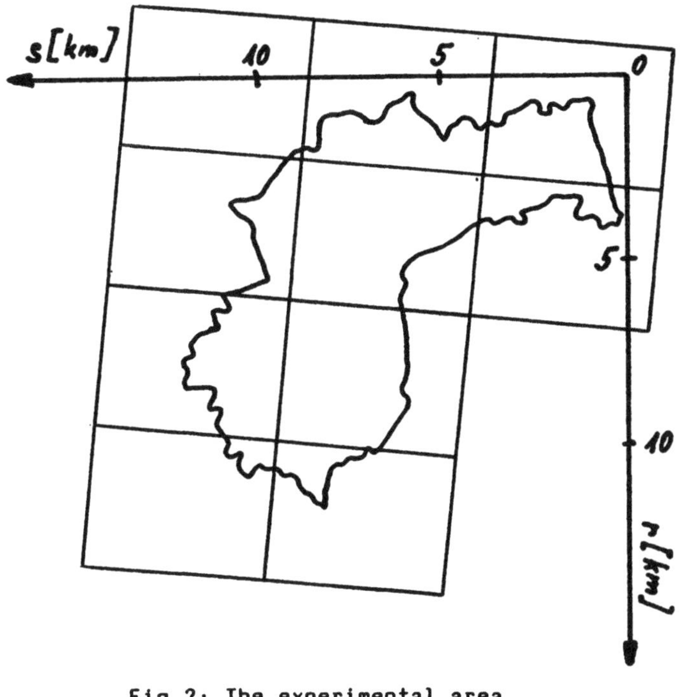

Fig.2: The experimental area

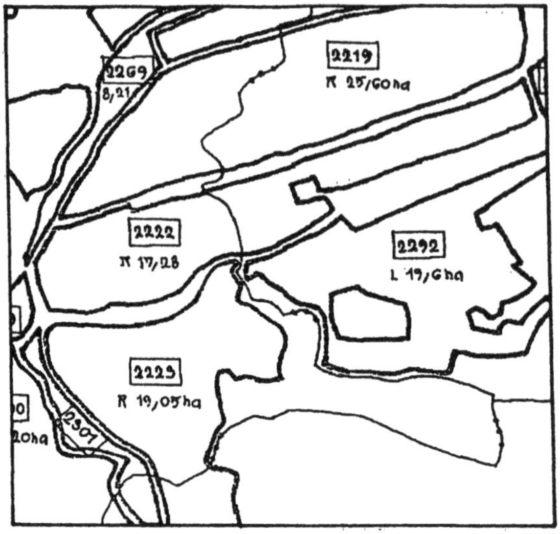

Fig.3: The agriculture land layer

Fig.4: The result of processing of the agriculture land layer

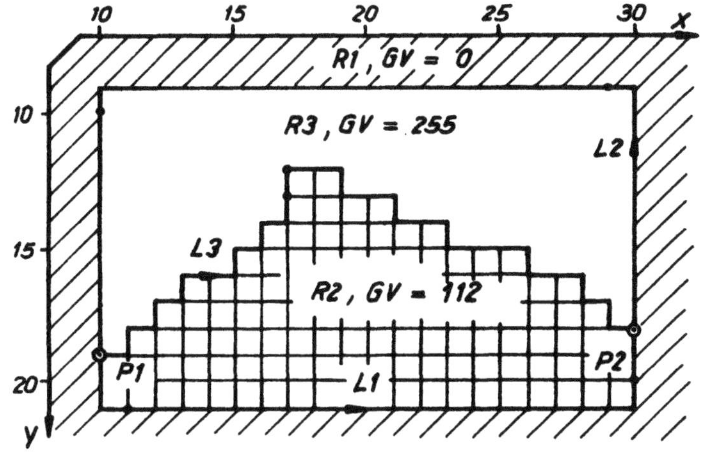

a)

P:	No	coordinates x	coordinates y	incident lines east	south	west	north
	P1	10	19	-L3	-L1	0	+L2
	P2	30	18	0	+L1	+L3	-L2

L:	No	incident points begin	end	incident regions right	left	metric list begin-addr.	end-addr.
	L1	P1	P2	R1	R2	1	4
	L2	P2	P1	R1	R3	5	8
	L3	P1	P2	R2	R3	9	13

R:	No	label GV	boundary start line
	R1	0	+L1
	R2	112	+L3
	R3	255	-L3

M: address	1	2	3	4	5	6	7	8
coord. (x, y)	(10,19)	(11,21)	(30,20)	(30,18)	(30,18)	(29,9)	(10,10)	(10,19)
address	9	10	11	12	13			
(x, y)	(10,19)	(17,13)	(17,12)	(24,15)	(30,18)			

b)

Fig. 1 : Example of the cell list description
a) picture b) cell list
P-branching points, L-lines, R-regions, M-metric data, GV-gray value

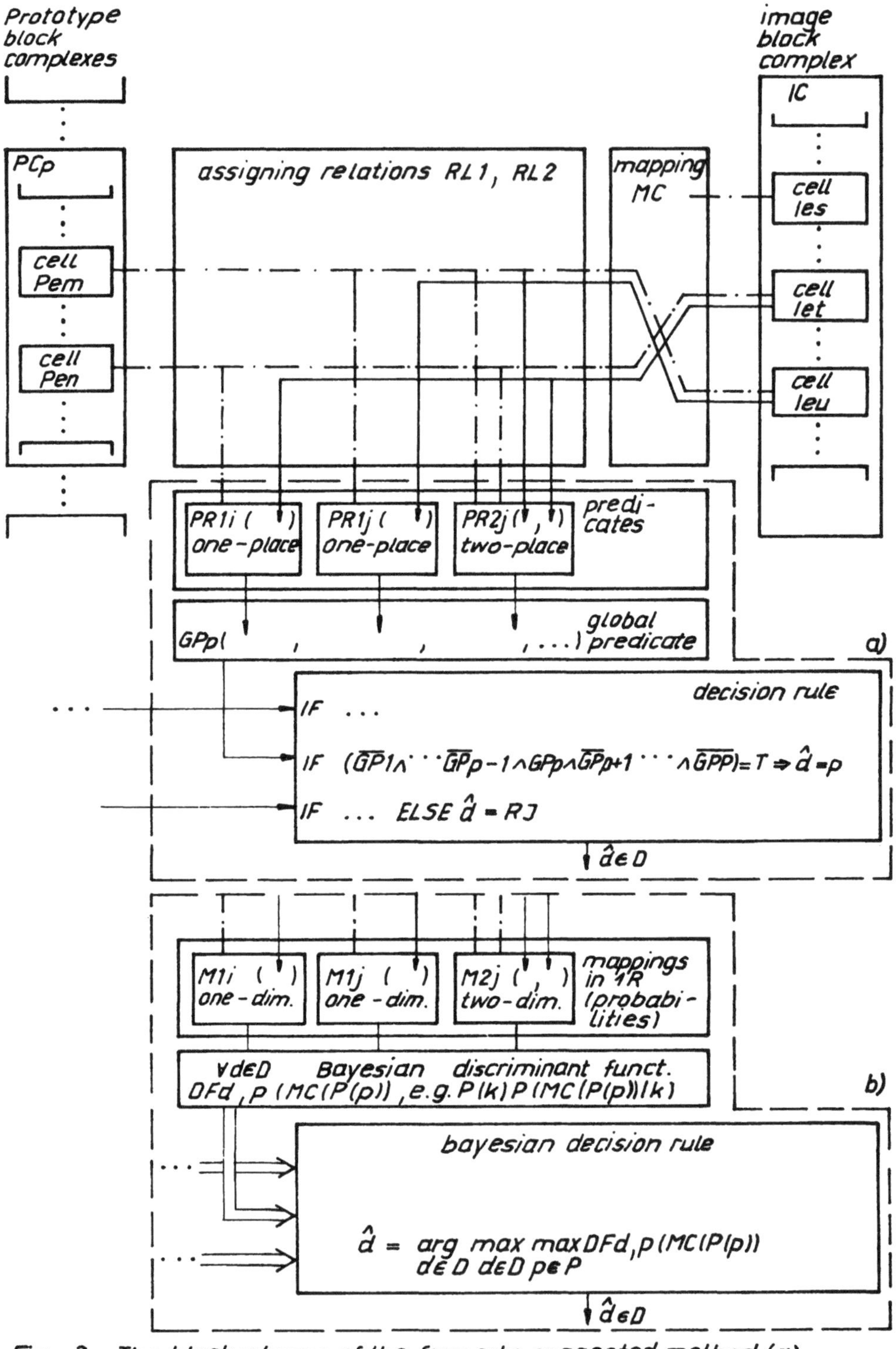

Fig. 2: The blockscheme of the formerly suggested method (a) and the generalized method (b) put by exchange in (a).

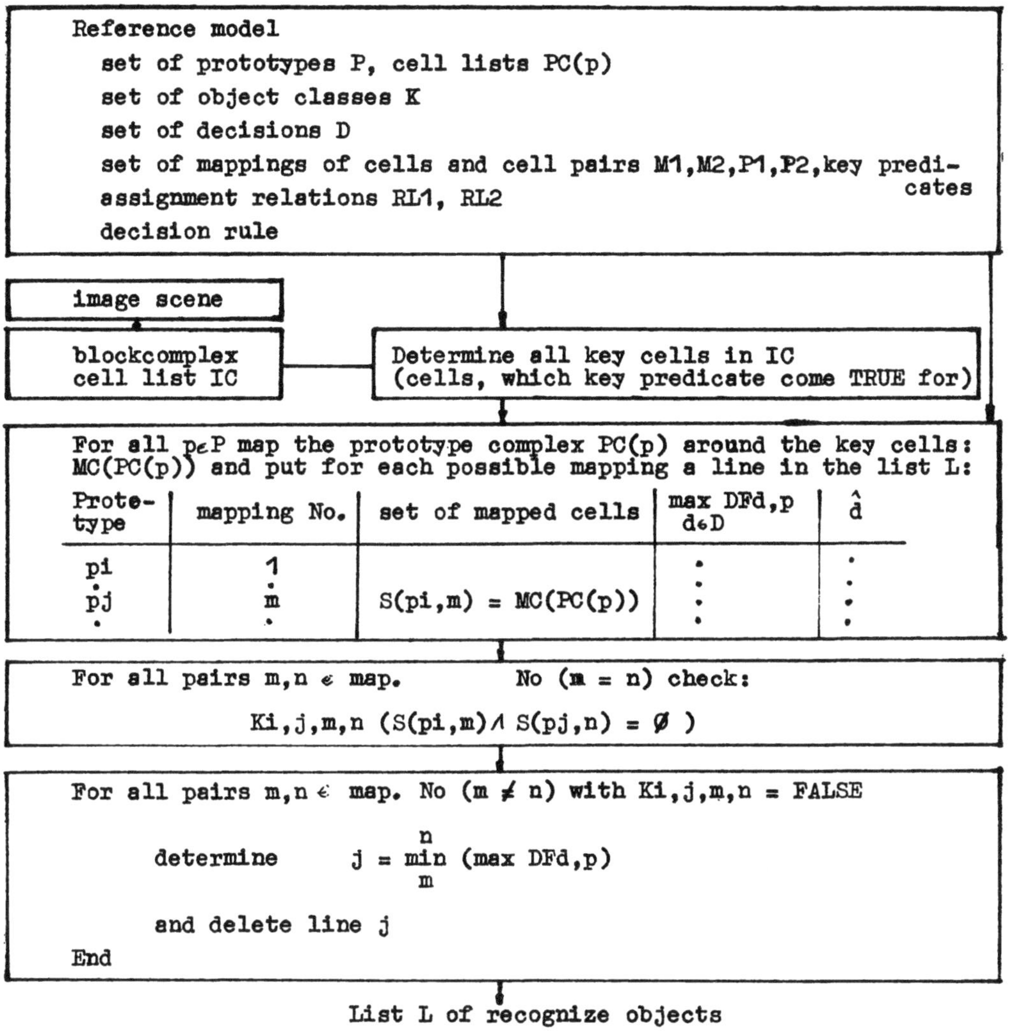

Fig. 3: Recognition scheme

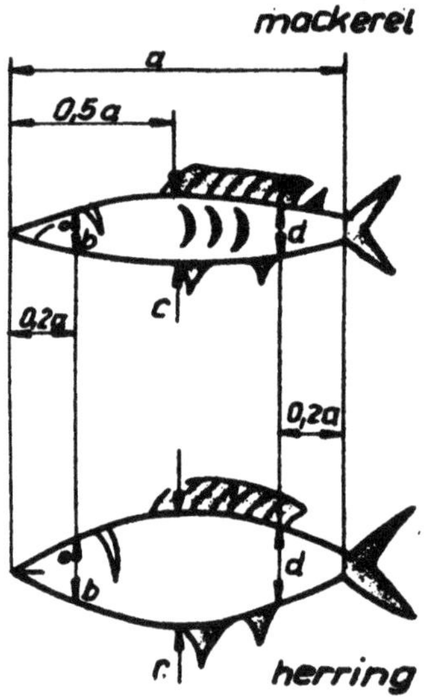

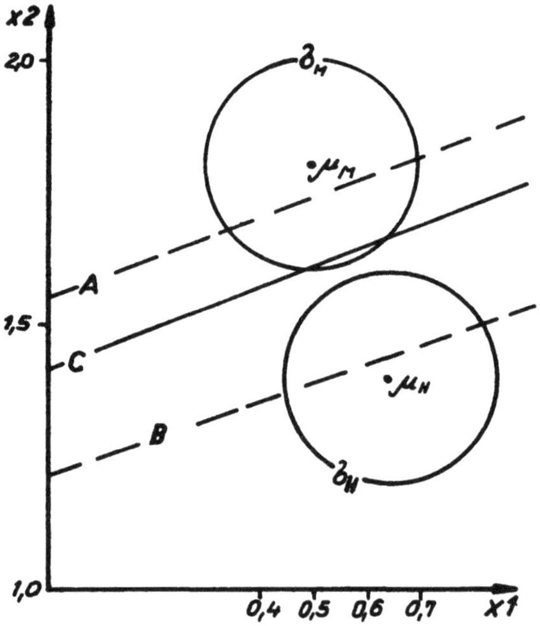

Fig. 4 : Pictures of the objects Fig. 5 : The feature space

Prototype image P1	Prototype image P2

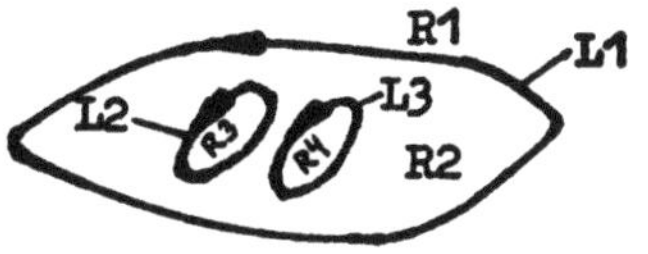

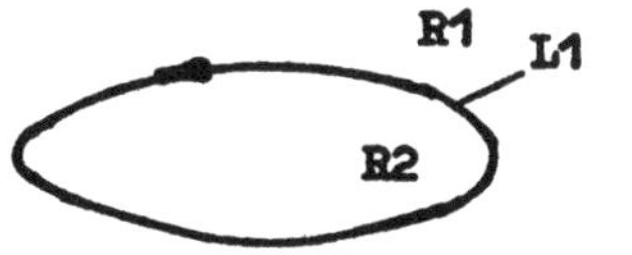

cell list PC 1 of the prototype P1

P:	-	- -	- - - -	
L:	L1 L2 L3	0 0 0 0 0 0	R1 R2 R2 R3 R2 R4	- - - - - -
R:	R1 R2 R3 R4	- - - -	L1 -L1 -L2 -L3	
M:	-			

cell list PC 2 of the protype P2

P:	-	- -	- - - -	
L:	L1	0 0	R1 R2	- -
R:	R1 R2	- -	L1 -L1	
M:	-			

Predicates and Mappings assigned to cells of PC 1/2

$M11(R) = \frac{1}{\sigma^2}\left|\underline{x}-\underline{\mu}_M\right|^2$ $\qquad$ $M12(R) = \frac{1}{\sigma^2}\left|\underline{x}-\underline{\mu}_H\right|^2$

M21(Ri,Rj) = lnP(STRIPE|KM) $\qquad$ PR11(R) = (R contains $\leq$2 regions)

M22(Ri,Rj) = lnP(STRIPE|KH) $\qquad$ PR21(R,R1) = (R imbedded in R1)

with STRIPE = TRUE

IF(SHAPE-PARAMETER(Ri)$>\varepsilon$)$\wedge$

(SHAPE-PARAMETER(Rj)$>\varepsilon$)$\wedge$ (Ri $\parallel$ Rj)

$PR10(R) = (S_{1U} \leq x_1 \leq S_{1O}) \wedge (S_{2U} \leq x_2 \leq S_{2O}) \wedge (a * c > S_3)$ $\qquad$ Si-thresholds

assignments RL1,RL2

	M11	M12	M21	M22	key PR10
R2	x	x			x
R3			x	x	x
R4			x	x	x

	M11	M12	PR11	PR21	key PR10
R1				x	x
R2	x	x	x	x	x

discriminant functions

IF PR10(MC(R2)) = TRUE THEN

$DF_{KM,P1} = \ln P(KM) + M21 - M11$

$DF_{KH,P1} = \ln P(KH) + M22 - M12$

ELSE $DF_{KM,P1} = DF_{KH,P1} = -\infty$

$D_{RJ,P1} = -\infty$

IF PR 10(MC(R2)) PR11 PR21=TRUE

$DF_{KM,P2} = \ln P(KM) + \ln P(STRIPE\ KM) - M11$

$DF_{KH,P2} = \ln P(KH) + \ln P(STRIPE\ KH) - M12$

ELSE $DF_{KM,P2} = DF_{KH,P2} = -\infty$

$D_{RJ,P2} = -\infty$

Fig. 6: Components of method 1

AUTOMATIC FEATURES GENERATION ALGORITHMS BASED ON THE SHAPE-DISCRIMINATING MAPPING

S.A.Kuz min[1]

1. Introduction

Nowadays great attention is paid to the topics associated with increasing the flexibility of machine vision systems intended for industrial applications, to ones of extention of their abilities in readjustment of the system from one task to another, and to improvement of their user-friendliness. The flexibility of contemporary commercially available industrial vision systems is achieved usually by increasing the number of different hardware and/or software modules, by providing the possibility of their more or less operative reconfiguration. The choice of adequate configuration for a particular application, however, turns out often to be a non-trivial task requiring the participation of a qualified specialist. As a typical example of such a situation one can refer to the important practical task of part sorting where a priori (on the stage of the vision system development) the features which should be used to discriminate the workpieces are unknown. An important peculiarity of such a problem is the fact that the system developer knows a priori what tranformation can an object image undergo. Particularly, if parts are placed on a working table over which a camera is fixed and the parts are non-overlapping then the transformation includes only rotations and translations. Other tasks may involve more complex trasformations (scaling, affine transforms, etc.).

Designing a computer vision system for such tasks one must develop a set of features which would be invariant to this transformation and provide desired discriminating power. The popular method of moment invariants [1] provides a formal procedure to construct a theoretically complete set of features invariant to a given class of transformations. In practice it does not allow, however, to obtain a sufficiently wide set because of the computational complexity of these features. Here we propose a method to generate local as well as global features, which provides a homogeneity of computations and therefore is more attractive from the point of view of practical implementations.

2. Local features

The method can be outlined as follows. Suppose that after a preprocessing of a digitized image an object contour was obtained. The contour is a set of points $S = \{x_i\}$, $i = 1,2..,K$ in plane (fig. 1), where K is the number of points in the contour. In this set let us pick out all possible overlapping subsets $\{s^N_j\}$, $j = 1,2,..., N!/K!(N-K)!$ including

---

[1] Institute of Control Sciences USSR 117806 Moscow Profsojuznaja 65.

each the fixed number of points N (fig.1). For such subsets and a known transformation G_p of the plane, where p is the vector of the transformation parameters, one can comparatively easily construct a set of so-called first integrals, using methods of linear differencial equations theory. These first integrals $\{u_m(x_1,x_2,\dots,x_N)\}$, $m = 1,\dots,2N-R$, where R is the number of mutually independent parameters $\{p_l\}$, $l = 1,2,\dots,R$ of the transformation, are invariant to this transformation and uniquelly define the "form" of any such a subset. In other words each subset can be uniquelly characterized by the vector of the values of these first integrals, which will be the same only for those subsets which can be completely matched using the transformation in hand. The outlook of the functions u depends only on the transformation G_p and the number N and is independent of the form of the set S. Moreover because of the first integrals properties any invariant defined on the subsets $\{s^N_j\}$ is coupled with them by a functional dependence. Therefore it is relevant to call the function u (vector one, generally speaking) as the universal invariant (UI). In the case of rotations and translations and N = 2 the UI is one-dimensional:

$$u = (x_1-x_2)^2 \qquad (1)$$

for N = 3:

$$\begin{aligned} u_1 &= (x_1-x_2, x_1-x_3) \\ u_2 &= [x_1-x_2, x_1-x_3] \\ u_3 &= (x_1-x_2)^2 \end{aligned} \qquad (2)$$

where x_i are the vectors of the set points coordinates in plane, and [*,*] and (*,*) are vector and dot products of vectors, correspondingly.

Statement 1. If the number N is such that

$$\text{rank} \begin{pmatrix} f_1(x_1) & f_1(x_2) & \dots & f_1(x_N) \\ f_2(x_1) & f_2(x_2) & \dots & f_2(x_N) \\ & & \cdots & \\ f_r(x_1) & f_r(x_2) & \dots & f_r(x_N) \end{pmatrix} = R$$

where $f_i(x)=\partial G_p x/\partial p_i|_{p=0}$ (the vector defining the magnitude and the direction of movement of a point in plane under small change of the parameter a_i of multivariate tranformation G_p in a neighbourhood of zero), R is the number of independent parameters of the transformation, then the UI $u(x_1,x_2,\dots,x_N)$ obtained for a subset consisting of N non-coinciding points and for the transformation G_p can be solved for the coordinates of any point presented in it.

Thus in the conditions formulated in the statement 1 knowing N-1 point of a subset and the corresponding values of the UI one can uniquelly determine the coordinates of the N-th point. The way of points indexing is insignificant. In other words the components of the UI form "rigid links" which tie a contour point to some configuration of (N-1) points. Particularily for the first N points of a contour one can compute the UI which would tie N-th point to the first N-1 points of the contour. After a shift by one point along the contour new values of the UI will be obtaind, tiyng (N+1)-th point to the previous (N-1) points. Accordingly, the two invariants obtaind tie (N+1)-th point to the first (N-1) points of the contour, and so on. Thus on passing a local contour segment an invariant description of the segment will be obtained which uniquelly characterizes its form [2].

When such a "window" of the width N moves along a contour the vector of the UI circumscribes a trajectory - the hodograph which on the one hand is an invariant signature of the contour and on the other hand is a result of a local mapping of contour segments of the length N into the space of the UI components. The properties of the UI are such that all the N-plets having the same form are mapped into the same point in the hodograph. In this sense the above mapping is similar to the Hough transform which maps segments of the same straight line into the same point.

To get a feel of the hodograph appearance some examples of it for simple geometrical figures are presented in figure 2 for two-dimensional invariant (2). (The third component for uniformely spaced contour points does not contain any information because in this case it is a constant.) The hodographs presented do not vary under traslations and rotations of objects and they may be used for selecting the unique local features, i.e. features the presence of which in an image permits to detrmine uniquelly the object type and position. To do this one must match hodographs corresponding to different objects (fig. 3) and for each hodograph find the segments (if any) which do not coincide with any segments of the other hodographs (thin lines in fig. 3). Here the methods of the cellular logic can be used. Because of the above mentioned properties of the UI one may say that if there are any unique features then they can be detected by the hodographs.

The above mapping enables to find out features which are presented in an image exactly a given number of times. Such features are often useful for structural methods of recognition. In this case one must compute not a hodograph which is simply a trace of the UI vector end, but a histogram (as in the Hough transform). The values of the histogram are the numbers of contour segments having the same form which corresponds to a given UI vector.

Using the UI for identification of local features opens one more perspective. Let an object has the form shown in fig. 4a. The "teeth" in this figure will be mapped to a cycle in hodograph (fig. 4b). Such hodograph cycles correspond to contour fragments whose local form is repeated periodically.

3. Global features

If one computes the histogram $H(u)$ of the UI values corresponding to all possible subsets $\{s_j\}$ of the contour set to be analyzed then a global description of the set will be obtained. The following statement is valid for this description.

Statement 2. For any set S there is a number $d>0$ such that under any deformations of S causing changes Δu in the UI vectors for subsets $\{s_j\}$ $(0<\Delta u<d$ but at least for one subset $u>0)$ the histogram $H(u)$ will change: $\|\Delta H(u)\|>0$.

In other words the histogram "feels" any deformations of a set of points. As global features one can use coefficients of an expansion of the histogram $H(u)$ in series by some functional basis defined on the space of UI vectors.

$$F_m = (q_m,H) = \sum_{x_i\subset S} q_m(u(x_1,x_2,\dots,x_N)) \qquad (3)$$

The main difficulty in constructing the feature system (3) is associated with determination of N and with deduction of the UI. This procedure, however, is formalized and must be executed only once for all the system of features simultaneously.

If the number N were chosen to be less then the value recommended by the statement 1 then the statement 2 would become invalid. The feature set (3), however, remains sufficiently wide and the computational efficiency of the resulting features will be higher as the multipleness of summation in eq. (3) is redused.

So returning to the case of parallel translations and rotations and using the invariant (1) for N=2 we obtain the features according to the eq. (3) as follows

$$J_m = \sum_{i=1}^{K} \sum_{j=1}^{K} q_m(x_i - x_j)^2 \qquad (4)$$

The manifestaton of the deviation from the requirement of the statement 1 is the additional invariancy of the features (4) to object mirroring.

In summary the proposed approach for generating invariant features for classification of descrete sets has the following advanteges:

1) homogeneity - the features differ from each other only by functions q which can be chosen computationally effective; this enables a compact hardware and/or software implementation of the features;

2) large dimensionality of the feature space, which enables to provide high discriminating power of classification module and to automate the procedure of choosing the minimal subset of features for classification of a particular nomenclature of objects;

3) universality - the features are not associated with any topological peculiarity of contours being analyzed.

4. Experiments and discussion

The global features (4) with the invariant (1) were used in the robot vision system intended for palletizing of disodered non-touching industrial parts of similar dimensions [3]. One of the tasks of the vision system was classification of parts to enable the robot to put the parts in appropriate pallets. The stage of classification of an object represented by its contour (100-200 points) under purely software computation of the features took about 0,3-0,9 s on LSI-11.

The hodographs were used in a more recent version of the robot vision system which was orientned to the task of classification and localization of touching and partially overlapping 2-D parts. The system includes two LSI-11. One of them was used for low level vision including computation of the local features and the other - for "intelligence" (hypothesis generation and verification, scene interpretation).

Computation of the hodograph for a contour containing about 100 points with the window of 10 points took approximately 90 ms. the dependance between the length of a contour and the time of hodograph computation was approximately linear. The window size has little influence on the time of analisys; the choice of this parameter is determined by a desired compromise between the stability of the discription in respect to image discretization noises and discriminating power of the resulting hodograph.

The experiments with different objects have shown high discrminating power of the features generated by the metodology proposed. Now the efforts are directed to creation of special-purpose hardware means for implementation of algoritms of feature computation. These means will enable real-time operation with complex transformations such as affine one and will make it possible to analize 3-D scenes in robotic applications.

5. References

[1] Hu M.K.: Visual pattern recognition by moment invariants. IEEE Trans. on Inform. Theory. IT-8 (1962), 179-187.

[2] Kuz min S.A., Il'in A.G.: Methods for generating the features for recognition of partially visible objects in computer vision systems. Mesurements, control, automation. (1982), 1, 56-77 (in russian).

[3] Petrov A.A., Kuz'min S.A.: Image analysis in a vision system for adaptive robots based on uniform algorithms. Computers and Artificial Intelligence, 8 (1989), 1, 93-103.

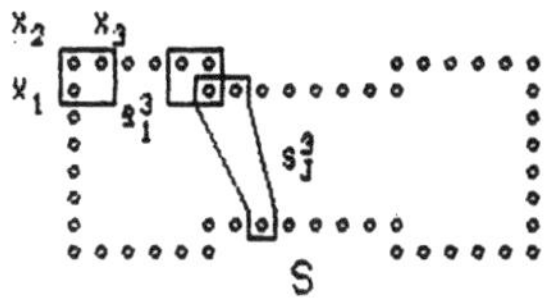

Fig. 1. Subdivision of a contour set into subsets for computation of UI.

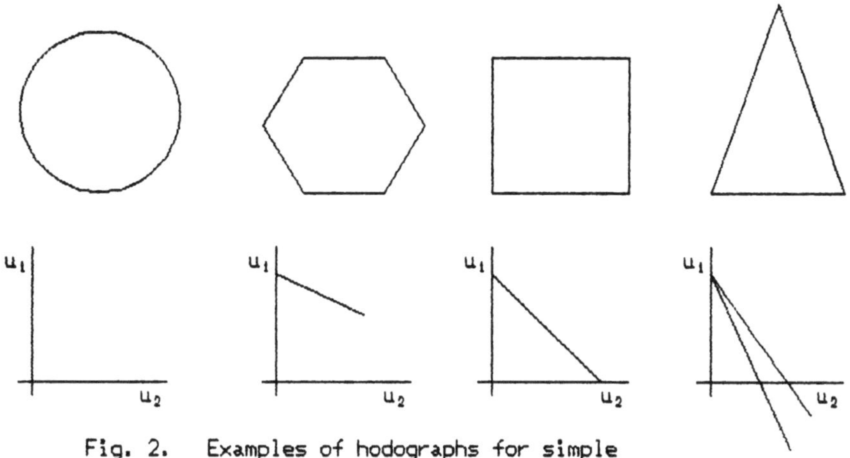

Fig. 2. Examples of hodographs for simple geometrical figures.

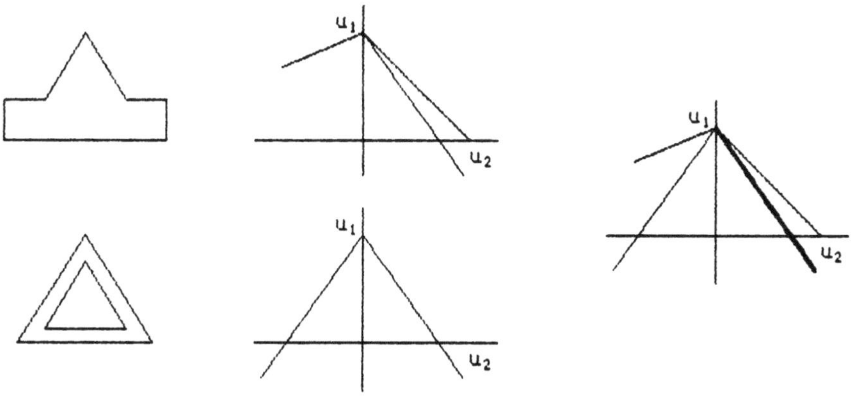

Fig. 3. Extracting unique features by matching hodographs.

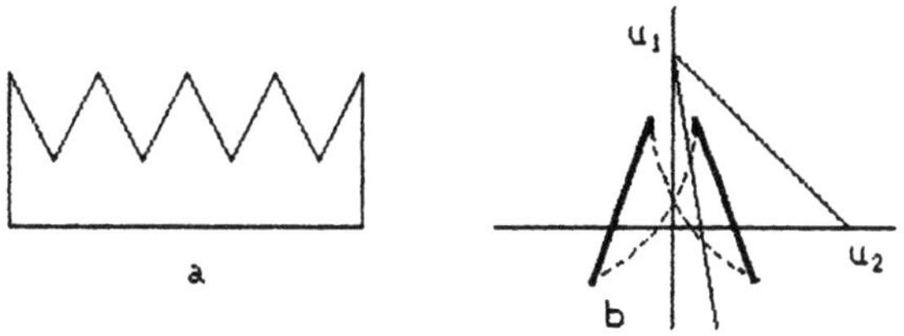

Fig. 4. Cycles in hodograph (b) indicating periodic contour structures.

Journal of New Generation Computer Systems

Журнал новых поколений вычислительных систем

Edited by order of the Coordination Council for Computer Technology and Informatics of the Academies of Sciences of the Socialist Countries at the Central Institute for Cybernetics and Information Processes of the Academy of Sciences of the GDR

Editor-in-Chief: Volker Kempe, Berlin
Assisted by an international Editorial Board

Journal of New Generation Computer Systems is devoted to all major activities in developing computer systems with new properties and novel application areas, i.e. to all efforts in creating new generation computer systems.
This journal reports on advances, experiences, and recent results in realizing national and international research and development programs and projects, in particular, co-operation projects of the socialist countries, all aiming at innovative computer systems. The major goal of the journal is to exchange information of the latest developments in hardware, software, and applications for processing knowledge and large amounts of data.
Considering these main characteristics the scope of the journal includes the following areas of interest:

— very large-scale integration
— new design and manifacture technologies for computer systems
— innovative architectures for various applications
— artificial intelligence-based software engineering
— knowledge processing
— man-machine communication
— theory and artificial intelligence research.

Besides these technical topics, the journal has columns on conference reports, calendar of events, book reviews etc.

Published yearly, 1 Volume (4 Numbers).
Volume 1, 1988.
17 cm by 24 cm. Annual list of contents.
Annual subscription DM 120,—; single number DM 30.
Order number: 1100. ISSN 0863-0445.

Orders can be sent

— *in the GDR:* to the Postzeitungsvertrieb or to the Akademie-Verlag Berlin, Leipziger Str. 3—4, PF. 1233, DDR-Berlin 1086;
in the other socialist countries: to a bookshop for foreign language literature or to the competent news-distributing agency;
— *in the FRG and Berlin (West):* to a bookshop or to the wholesale distributing agency Kunst und Wissen, Erich Bieber OHG, Wilhelmstraße 4—6, D-7000 Stuttgart 1;
— *in the other Western European countries:* to Kunst und Wissen, Erich Bieber GmbH, General Wille-Str. 4, CH-8002, Zürich;
— *in other countries:* to the international book- and journal-selling trade; to Buchexport, Volkseigener Außenhandelsbetrieb der DDR, Postfach 160, DDR-Leipzig 7010; or to the Akademie-Verlag Berlin, Leipziger Str. 3—4, PF. 1233, DDR-Berlin 1086.

AKADEMIE VERLAG BERLIN
THEORIA CUM PRAXI
Deutsche Demokratische Republik
DDR-1086 Berlin, Leipziger Strasse 3-4 Postfach 1233

Theoretische Grundlagen der Künstlichen Intelligenz

Herausgegeben von
JAN GRABOWSKI /
KLAUS P. JANTKE /
HELMUT THIELE

(Informatik – Kybernetik – Rechentechnik, Band 31)

1989, ca. 260 Seiten – 17 × 24 cm
DDR ca. 40,– M; Ausland ca. 48,– DM
ISBN 3-05-500 601-1
Bestell-Nr. 763 910 0
Bestellwort: Kuenstl. Intelligenz 9157

Dieser Sammelband vermittelt eine Übersicht über das sich in stürmischer Entwicklung befindliche Gebiet der Künstlichen Intelligenz und kann auch als Einführung in diese Thematik dienen. Behandelt werden Konzepte der Wissensrepräsentation, automatisches Theorembeweisen, Expertensysteme, logische Programmierung, automatische Programmsynthese, Wissensbanken und Datenbanken sowie weitere aktuelle Themen der KI-Forschung.

Interessentenkreis: Informatiker, Mathematiker, Sprachwissenschaftler

---

NIKOLAY PETKOV

Systolische Algorithmen und Arrays

(Informatik – Kybernetik – Rechentechnik, Band 27)

1989, ca. 300 Seiten – ca. 80 Abbildungen
17 × 24 cm – DDR ca. 45,– M;
Ausland ca. 55,– DM
ISBN 3-05-500 661-5
Bestell-Nr. 763 972 5
Bestellwort: Petkov, Syst. Algorith. 9199

Das Buch bietet eine Einführung in die systolischen Algorithmen und Arrays, in den letzten Jahren mit den Erfolgen der VLSI-Technik entstanden und zu einer der wichtigsten Entwicklungsrichtungen der parallelen Datenverarbeitung herangewachsen ist. Es wird auf die Vorteile der systolischen Arrays für den Entwurf von VLSI-Schaltkreisen sowie den Zusammenhang mit anderen Rechenmodellen für Parallelverarbeitung, wie SIMD-, MIMD-, Datenfluß-Computer u. a. eingegangen. Der Autor veranschaulicht das systolische Konzept an zahlreichen Beispielen aus der numerischen Mathematik, der digitalen Signal- und Bildverarbeitung sowie der algorithmischen Geometrie.

Interessentenkreis: Hoch- und Fachschulingenieure der Rechentechnik und des Schaltkreisentwurfs

---

MARTIN ENGELIEN /
HEINZ STAHN

Software Engineering

CAMARS-Technologie

1989, 288 Seiten – 154 Abbildungen
14,5 × 21 cm – DDR ca. 36,– M;
Ausland ca. 45,– DM
ISBN 3-05-500 653-4
Bestell-Nr. 763 957 3
Bestellwort: Engelien/Stahn 9197

Die Autoren erläutern die von ihnen entwickelte CAMARS-Technologie und ihre Anwendung für den Entwurf und für die Implementation von Software für Steuerungen im Computer Aided Manufacturing und zeigen deren hohe Produktivität in zahlreichen Beispielen.

Interessentenkreis: Hoch- und Fachschulingenieure der Datenverarbeitung, Informatiker

KLAUS VOSS

Theoretische Grundlagen der digitalen Bildverarbeitung

(Informatik – Kybernetik – Rechentechnik, Band 23)

1988, 159 Seiten – 68 Abbildungen
13 Tabellen – 17 × 24 cm – Lederin
DDR 30,– M; Ausland 40,– DM
ISBN 3-05-500 513-9
Bestell-Nr. 763 846 6
Bestellwort: Voss, Dig. Bildverarb. 9124

Der Autor zeigt, daß diskrete Topologie, diskrete Geometrie und diskrete Funktionstheorie das mathematische Handwerkszeug der digitalen Bildverarbeitung bilden. Die Topologie umfaßt den größten Teil des Buches, das mit vielen Abbildungen, Übungsaufgaben und Randbemerkungen versehen ist und eine systematische Einführung für den Anfänger sowie eine anregende Lektüre für den Fachmann der digitalen Bildverarbeitung bietet.

Interessentenkreis: Hoch- und Fachschulingenieure der Bildverarbeitung und Datenerkennung

---

MICHAEL GÖSSEL / BURGHARD REBEL / REINER CREUTZBURG

Speicherarchitektur und Parallelzugriff

(Informatik – Kybernetik – Rechentechnik, Band 28)

1989, 219 Seiten – 87 Abbildungen
22 Tabellen – 17 × 24 cm
DDR ca. 30,– M; Ausland ca. 38,– DM
ISBN 3-05-500 599-6
Bestell-Nr. 763 907 1
Bestellwort: Goessel, u. a., Speicher 9155

Mit diesem Buch liegt die erste zusammenfassende Darstellung einer einheitlichen Theorie paralleler Speicher vor. Dem Ingenieur zeigt es, wie parallele Speicher funktionieren und entworfen werden. Es betrifft die Gebiete Bild- und Signalverarbeitung, Computergraphik, Supercomputer sowie Parallelverarbeitung.

Interessentenkreis: Theoretisch interessierte Informatiker

HANS BANDEMER / SIEGFRIED GOTTWALD

Einführung in FUZZY-Methoden

Theorie und Anwendung unscharfer Mengen

(Wissenschaftliche Taschenbücher, Reihe Mathematik/Physik)

1989, ca. 210 Seiten – 11 Abbildungen
3 Tabellen – 12,5 × 20 cm,
DDR 16,– M; Ausland 18,– DM
ISBN 3-05-500 554-6
Bestell-Nr. 763 871 4
Bestellwort Bandemer/Gottwald 7305

Die Autoren stellen eine neue Möglichkeit der Beschreibung mathematischer Sachverhalte vor: die graduelle Zugehörigkeit von Elementen zu einer Menge. Sie erläutern grundlegende Begriffe und Ergebnisse der Theorie unscharfer Mengen und zeigen die vielfältige Anwendbarkeit dieser Theorie auf konkrete technische, medizinische, soziologische und ökonomische Probleme.

Interessentenkreis: Mathematiker, Naturwissenschaftler, Ingenieure, Ökonomen

Band 1

Inverse and Improperly Posed Problems in Differential Equations

Proceedings of the Conference on Mathematical and Numerical Methods, held in Halle/Saale (GDR) from May 29 to June 2, 1979

Edited by Gottfried Anger

296 Seiten, 1979 (vergriffen)

Band 2

Fundamentals of Computation Theory FCT '79

Proceedings of the Conference on Algebraic, Arithmetic and Categorial Methods in Computation Theory, held in Berlin/Wendisch-Rietz (GDR) September 17–21, 1979

Edited by Lothar Budach

576 Seiten, 1979

Band 3

Decidability and Generalized Quantifiers

By Andreas Baudisch, Detlef Seese, Hans-Peter Tuschik, and Martin Weese

XII, 235 Seiten, 1980

Band 4

Zur Elastizitäts- und Thermoelastizitätstheorie

Potentialtheoretische Methoden in Anwendung auf verschiedene Modelle

Von Lothar Jentsch und Johannes Maul

157 Seiten, 1980

Band 5

Systems Analysis and Simulation 1980

Proceedings of the International Symposium held at Berlin (GDR) September 1–5, 1980

Edited by Achim Sydow

496 Seiten, 1980 (vergriffen)

Band 6

Nichtlineare diskrete Optimierung

von Eberhard Girlich und Michail M. Kowaljow

218 Seiten, 1981

Band 7

Completeness Problems in the Structural Theory of Automata

By Jürgen Dassow

148 Seiten, 1981

Band 8

Darstellung von Lösungen linearer elliptischer Differentialgleichungen

Von Günther Wildenhain

92 Seiten, 1981

Band 9

Introduction to the Spectral Theory of Operators in Spaces with an Indefinite Metric

By I. S. Iohvidov, M. G. Krein, and H. Langer

120 Seiten, 1982 (vergriffen)

Band 10

Nonlinear Time-discrete Systems

A General Approach by Nonlinear Superposition

By Michael Gössel

115 Seiten, 1982 (vergriffen)

Band 11

Preprimal Algebras

By Klaus Denecke

162 Seiten, 1982

Band 12

Ergodic Theory and Related Topics

Proceedings of the Conference held in Vitte/Hiddensee (GDR) October 19–23, 1981

Edited by Horst Michel

232 Seiten, 1982

Band 13

Singularly Perturbed Differential Equations

By H. Goering, A. Felgenhauer, G. Lube, H.-G. Roos, and L. Tobiska

176 Seiten, 1983

Band 14

Leben wir in einer Volterra-Welt?

Ein ökologischer Zugang zur angewandten Systemanalyse

Von Manfred Peschel und Werner Mende

233 Seiten, 1983

Band 15

Markov Decision Problems with Countable State Spaces

Optimality Criteria – Algorithms – Clustering

By Hans Michael Dietz and Volker Nollau

174 Seiten, 1983

Band 16

Stochastische Verfahren zur Suche nach dem Minimum einer Funktion

Von Ryszard Zieliński und Peter Neumann

133 Seiten, 1983

Band 17

Invertibility and Asymptotics of Toeplitz Matrices

By Albrecht Böttcher and Bernd Silbermann

200 Seiten, 1983

Band 18

Structural Induction on Partial Algebras

Introduction to Theory and Application of Partial Algebras – Part II

By Horst Reichel

205 pages, 1984

Band 19

Algebraic Methods for Toeplitz-like Matrices and Operators

By Georg Heinig and Karla Rost

212 pages, 1984

Band 20

Frege Conference 1984

Proceedings of the International Conference held at Schwerin (GDR), September 10–14, 1984

Edited by Gerd Wechsung

408 pages, 1984

Band 21

Numerical Analysis of Selected Semilinear Differential Equations

Edited by Christian Großmann, Thomas Riedrich, and Hartmut Schönheinz

203 pages, 1984

Band 22

Singular Ordinary Differential Operators and Pseudodifferential Equations

By Johannes Elschner

200 pages, 1985

Band 23

Lotka-Volterra-Approach to Cooperation and Competition in Dynamic Systems

Proceedings of the 5th Meeting of UNESCO's Working Group on Systems Theory held on the Wartburg, Eisenach (GDR) at March 5–9, 1984

Edited by Werner Ebeling and Manfred Peschel

271 pages, 1985

Band 24

Convergence Structures 1984

Proceedings of the Conference on Convergence held in Bechyně (Czechoslovakia), September 24–28, 1984

Edited by Josef Novák, Werner Gähler, Horst Herrlich, and Jan Mikusiński

254 pages, 1985

Band 25

Parcella '84

Proceedings of the II. International Workshop on Parallel Processing by Cellular Automata and Arrays held in Berlin (GDR), September 25–27, 1984

Edited by Wolfgang Händler, Tamás Legendi, and Gottfried Wolf

239 pages, 1985 **(vergriffen)**

Band 26

Multiobjective and Stochastic Optimization Based on Parametric Optimization

By Jürgen Guddat, Francisco Guerra Vasquez, Klaus Tammer, and Klaus Wendler

175 pages, 1985

Band 27/28

Systems Analysis and Simulation 1985
I: Theory and Foundations
II: Applications

Proceedings of the International Symposium held in Berlin (GDR), August 26–31, 1985

Edited by Achim Sydow, Manfred Thoma, and Robert Vichnevetsky

per volume 495 pages, 1985

Band 29

Parcella '86

Proceedings of the III. International Workshop on Parallel Processing by Cellular Automata and Arrays held in Berlin (GDR), September 9–11, 1986

Edited by Tamás Legendi, Dennis Parkinson, Roland Vollmar, and Gottfried Wolf

279 pages, 1986

Band 30

Fuzzy Sets Applications, Methodological Approaches, and Results

Proceedings of the International Workshop on Fuzzy Sets Applications cosponsored by IIASA and convened in cooperation with IAEA held on the Wartburg, Eisenach (GDR), at March 3–8, 1985

Edited by Steffen Bocklisch, Sergei Orlovski, Manfred Peschel, and Yasushi Nishiwaki

233 pages, 1986

Band 31

Mathematical Methods of Specification and Synthesis of Software Systems '85

Proceedings of the International Spring School, held in Wendisch-Rietz (GDR), April 22–26, 1985

Edited by Wolfgang Bibel and Klaus P. Jantke

245 pages, 1986 (vergriffen)

Band 32

A Model Theoretic Oriented Approach to Partial Algebras

Introduction to Theory and Application of Partial Algebras – Part I

By Peter Burmeister

319 pages, 1986

Band 33

Finite Difference Method on Irregular Networks

A Generalized Approach to Elliptic Problems of Second Order

By Bernd Heinrich

206 pages, 1986

Band 34

Modern Algorithms for Large Sparse Eigenvalue Problems

By Arnd Meyer

125 pages, 1987

Band 35

Parametric Optimization and Related Topics

Edited by Jürgen Guddat, Hubertus Th. Jongen, Bernd Kummer, and František Nožička

407 pages, 1987

Band 36

Discretization in Differential Equations and Enclosures

Edited by Ernst Adams, Rainer Ansorge, Christian Großmann, and Hans-Görg Roos

250 pages, 1987

Band 37

Factorization of Measurable Matrix Functions

By Georgi Semenovich Litvinchuk and Ilya Matveyevich Spitkovskiî

372 pages, 1987

Band 38

Parallel Algorithms and Architectures

Proceedings of the International Workshop on Parallel Algorithms and Architectures held in Suhl (GDR), May 25–30, 1987

Edited by Andreas Albrecht, Hermann Jung, and Kurt Mehlhorn

205 pages, 1987

Band 39

Parametric Integer Optimization

By Bernd Bank and Reinhard Mandel

138 pages, 1988

Band 40

Computer Analysis of Images and Patterns 1987

Proceedings of the II. International Conference CAIP '87 on Automatic Image Processing, held in Wismar (GDR), September 2–4, 1987

Edited by Leonid P. Jaroslavskiî, Azriel Rosenfeld, and Wolfgang Wilhelmi

282 pages, 1987

Band 41

Multivariable Control

A Graph-theoretic Approach

By Kurt J. Reinschke

274 pages, 1988

Band 42

Equilibrium Distributions of Branching Processes

By Alfred Liemant, Klaus Matthes, and Anton Wakolbinger

240 pages, 1988

Band 43

Andreotti-Grauert Theory by Integral Formulas

By Gennadi Markovich Henkin and Jürgen Leiterer

270 pages, 1988

Band 44

Continual Means and Boundary Value Problems in Function Spaces

By Efim Mikhailovich Polishchuk

164 pages, 1988

Band 45

Advances in Mathematical Optimization

Invited papers dedicated to Prof. Dr. Dr. h. c. F. Nožička on occasion of his 70th birthday

Edited by Jürgen Guddat et al.

332 pages, 1988

Band 46/47

Systems Analysis and Simulation 1988

Proceedings of the International Symposium held in Berlin (GDR),

September 12–16, 1988

I: Theory and Foundations
II: Applications

Edited by Achim Sydow, Spyros G. Tzafestas, and Robert Vichnevetsky

426/425 pages, 1988

Band 48

Parcella '88

Proceedings of the IV. International Workshop on Parallel Processing by Cellular Automata and Arrays, held in Berlin (GDR),

October 17–21, 1988

Edited by Gottfried Wolf, Tamás Legendi, and Udo Schendel

379 pages, 1988

Band 49

Algebraic and Logic Programming

Proceedings of an International Workshop held in Gaussig (GDR), November 14–18, 1988

Edited by Jan Grabowski, Pierre Lescanne, and Wolfgang Wechler

278 pages, 1988

Band 50

Asymptotics for Elliptic Mixed Boundary Problems

Pseudo-Differential and Mellin Operators in Spaces with Conormal Singularity

By Stephan Rempel and Bert-Wolfgang Schulze

418 pages, 1989

Band 51

Geobild '89

Proceedings of the 4th Workshop on Geometrical Problems of Image Processing held in Georgenthal (GDR), March 13–17, 1989

Edited by Albrecht Hübler, Werner Nagel, Brian David Ripley, and Günter Werner

198 pages, 1989

Band 52

Splines in Numerical Analysis

Contributions to the International Seminar ISAM '89

held in Weissig (GDR), April 24–28, 1989

Edited by Jochen W. Schmidt and Helmut Späth

197 pages, 1989

Band 53

Complex Methods on Partial Differential Equations

Aspects in Complex Analysis

Edited by Claudio I. Withalm

258 pages, 1989

The processing of pictorial signals and the automatic analysis of images and patterns is one of the main application fields of information processing in general. Special applications cover the field from astronomy, biology, material research etc. to robot control, industrial inspection and interpretation of moving scenes.
Image processing has a strong influence on new computer generations (expert systems, array processors, artificial intelligence) and is strong influenced by new computers, devices, and sensors.
Image processing has an own theoretical framework for development of effective algorithms (discrete geometry, discrete topology, integral transformations etc.) and uses new principles in software systems.
Reader will find selected papers to the international conference CAIP '89 organized by WGMA concerning the above mentioned topics.

Die Verarbeitung von Bildsignalen und die automatische Analyse von Bildern und Mustern ist eins der Hauptanwendungsgebiete der Informationsverarbeitung. Spezielle Anwendungen reichen von Astronomie, Biologie, Materialforschung usw. bis zur Robotersteuerung, der industriellen Inspektion und der Interpretation bewegter Szenen.
Die digitale Bildverarbeitung hat einen starken Einfluß auf neue Computerarchitekturen (Expertensysteme, Array-Prozessoren, künstliche Intelligenz) und wird selbst stark beeinflußt durch neuartige Computer, Geräte und Sensoren.
Die digitale Bildverarbeitung gründet sich auf theoretische Modelle für die Entwicklung effektiver Algorithmen (diskrete Geometrie, diskrete Topologie, Integraltransformationen usw.) und nutzt neuartige Prinzipien von Softwaresystemen.
Der Leser findet in diesem Band ausgewählte Beiträge der Internationalen Konferenz CAIP '89, die von der WGMA in der KdT der DDR veranstaltet wurde.

ISBN 3-05-500755-7
ISSN 0138-3019

Bei Fragen zur Produktsicherheit wenden Sie sich bitte an:
If you have any questions regarding product safety,
please contact:

Walter de Gruyter GmbH
Genthiner Straße 13
10785 Berlin
productsafety@degruyterbrill.com